ELECTROMAGNETIC
WAVES

ELECTROMAGNETIC
WAVES

Microwave components and devices

R.G. Carter

Department of Engineering,
University of Lancaster

CHAPMAN AND HALL

LONDON · NEW YORK · TOKYO · MELBOURNE · MADRAS

UK	Chapman and Hall, 11 New Fetter Lane, London EC4P 4EE
USA	Van Nostrand Reinhold, 115 5th Avenue, New York NY10003
JAPAN	Chapman and Hall Japan, Thomson Publishing Japan, Hirakawacho Nemoto Building, 7F, 1–7–11 Hirakawa-cho, Chiyoda-ku, Tokyo 102
AUSTRALIA	Chapman and Hall Australia, Thomas Nelson Australia, 480 La Trobe Street, PO Box 4725, Melbourne 3000
INDIA	Chapman and Hall India, R. Sheshadri, 32 Second Main Road, CIT East, Madras 600 035

First edition 1990

© 1990 R.G. Carter

Typeset in 10pt Times by Best-set Typesetter Ltd
Printed in Great Britain by TJ Press (Padstow) Ltd, Padstow, Cornwall

ISBN 0 412 34190 5 (PB)   0 442 31183 4 (USA)

British Library Cataloguing in Publication Data
Carter, R. G. (Richard G)
  Electromagnetic waves.
  1. Electromagnetism
  I. Title
  537

ISBN 0–412–34190–5

Library of Congress Cataloging-in-Publication Data
Carter, R.G. (Richard Geoffrey)
    Electromagnetic waves : microwave components and devices / R.G. Carter.
      p.   cm.
    Includes bibliographical references.
    ISBN 0–442–31183–4
    1. Microwave devices   2. Electromagnetic waves.   I. Title.
  TK7876.C39   1990
    621.381′31—dc20   89-48148
                                                               CIP

To my father
Geoffrey William Carter
(1909–89)
A distinguished teacher of electromagnetic theory

Contents

Preface

Electromagnetic theory is fundamental to the whole of electrical and electronic engineering. As such it should surely be an essential part of the professional knowledge of all who call themselves electronic engineers. Yet it is a common complaint among teachers of the subject that students cannot be persuaded to take it seriously perhaps because of their obsession with digital electronics. It seems to me that this is very regrettable. Advances in high-speed digital electronics and in opto-electronics will present problems which cannot be solved without an understanding of electromagnetic theory. The EEC directive on electromagnetic compatibility to be adopted in 1992 will likewise demand a knowledge of fundamental principles. For these reasons it is vital that all students of electrical and electronic engineering should gain a basic knowledge of electromagnetics.

My own experience of grappling with problems in the engineering applications of electromagnetic theory has convinced me that the subject is usually presented in an over-mathematical way. This may be an additional reason why students find it unattractive. In this book I have introduced the subject in a physical and intuitive way making use of elementary mathematics for the most part. The emphasis is on the physical understanding which is the basis for solving problems. Those who eventually need to understand the full mathematical treatment will find that this book provides a good starting point. More often these days computer packages are used to solve electromagnetic field problems with complex boundary conditions.

Engineers generally prefer to work with circuit theory than with field theory. This is typified by the use of 'j notation' to extend the methods of analysis of d.c. circuits to a.c. problems. Microwave engineers normally work with transmission line equivalent circuits whenever possible. A major concern in this book is with the ways in which these equivalent circuits are developed.

The content of the book may be divided into three parts. In Chapters 1 to 4 the emphasis is on basic properties of electromagnetic waves. Chapters 5 to 10 deal systematically with the applications of the theory to a wide range of components and devices. Many of the applications are in microwave engineering but optical and e.m.c. topics are included wherever appropriate. Finally, Chapters 11 and 12 discuss microwave and e.m.c. measuring techniques and provide an overview of the applications of electromagnetic waves in a variety of systems. The aim throughout has been to provide the reader with the basic knowledge which will make the professional literature of the subject accessible. An extensive list of references is provided for this purpose. A small number of exercises are grouped at the end of each chapter (except Chapter 9 where they seemed inappropriate). These are mostly very straightforward. Their purpose is to help the reader to understand the main points in the text, to give confidence in handling the ideas and to give a feel for the numbers involved.

This book carries on the development of the subject from the point I reached in *Electromagnetism for Electronic Engineers* (1986). Those who have found that book helpful will find that the approach in this one is familiar. I hope that it will be found useful not only by students but also by those who discover later in their careers a need for a knowledge of electromagnetic theory.

I am indebted to a number of people who have influenced my own understanding of electromagnetism. My father taught me the value of 'thinking from first principles'. As an undergraduate I relied heavily on Bleaney and Bleaney (1976) which is a model of elegant simplicity in its treatment of the subject. I hope I may have achieved for engineers what that book did for students of physics. My present head of department, Colin Hannaford, was the first to draw my attention to the value of equivalent circuit methods. Finally, Schelkunoff (1943) introduced me to the idea of using transmission-line methods for electromagnetic wave problems.

In the writing of this book I have been heavily indebted to a number of people. Dr L.G. Ripley of the University of Sussex kindly commented on the manuscript and made many helpful suggestions. Dominic Recaldin and his staff at the publishers were patient beyond belief with an author who had a chronic inability to meet deadlines. Most of all I must thank my wife and family who had to put up with my frequent and lengthy disappearances into my study.

R.G. Carter
1989

Electromagnetic waves

1

1.1 INTRODUCTION

This book is about electromagnetic waves, the spectrum of radiation which ranges from the longest radio waves through the infrared and optical regions and on to hard X-rays and gamma rays. Figure 1.1 shows the electromagnetic spectrum and some of its uses. Engineers make use of every part of this vast range of frequencies in information systems of all kinds. Although it is sometimes possible to work with the systems without knowledge of the underlying physical principles there are occasions when this ignorance is a handicap. The whole subject is based on just four physical laws and the consequences of their application to problems with different boundary conditions. Many of these problems can be studied without the use of advanced mathematical methods. Indeed the use of those methods can hinder the growth of the physical understanding which really solves problems. The aim of this book is three-fold: to help the reader to gain a physical understanding of problems involving electromagnetic waves, to relate that understanding particularly to modern problems, and to provide a route into the professional literature of the subject.

Engineers first became aware of electromagnetic waves in the middle of the nineteenth century with the development of the electric telegraph. This was understood, however, in terms of circuit theory rather than electromagnetic field theory. At that time optics was regarded as a separate branch of physics. The work of James Clerk Maxwell, published in his Treatise on Electricity and Magnetism in 1873, provided for the first time a field theory of electromagnetic waves and evidence that light is also an electromagnetic phenomenon. The subsequent exploitation of that theory in radio, radar, television, satellite communications and coherent optics has produced a transformation of human life which is still going on. Maxwell's equations are therefore the starting point of any discussion of these subjects and also of the unwanted electromagnetic coupling which is of increasing concern to electronic engineers.

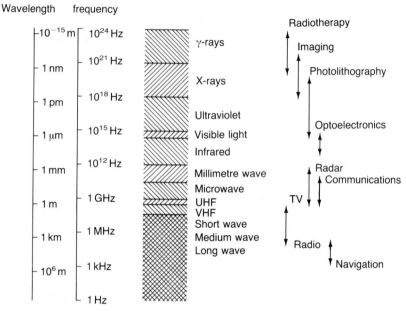

Fig. 1.1 Chart of the electromagnetic spectrum showing some of the uses to which electromagnetic waves are put.

1.2 MAXWELL'S EQUATIONS

The equations which are now known as Maxwell's equations are actually a summary of the four basic laws of electromagnetism (Carter, 1986 p. 12). These are listed below.

1. Gauss' theorem of electrostatics states that the flux of the electric flux density D, sometimes known as the electric displacement, out of a closed surface is equal to the total free charge enclosed.

$$\oiint D \cdot dA = \iiint \varrho \, dv, \tag{1.1}$$

where ϱ is the charge density and dA and dv are elements of the area of the surface and of its volume.

2. Gauss' theorem of magnetostatics states that the flux of the magnetic flux density B out of a closed surface is zero (because magnetic monopoles do not exist).

$$\oiint B \cdot dA = 0. \tag{1.2}$$

3. The magnetic circuit law as modified by Maxwell to include the displacement current (sometimes known as Ampère's law) states that the line integral of the magnetic field vector H around a closed path is equal to the total current flux (conduction plus displacement current) through a surface bounded by that path.

$$\oint H \cdot \mathrm{d}l = \iint \left(J + \frac{\partial D}{\partial t} \right) \cdot \mathrm{d}A, \tag{1.3}$$

where J is the conduction current density.

4. Faraday's law of electromagnetic induction states that the line integral of the electric field vector E around a closed path is equal to the rate of change of the magnetic flux through a surface bounded by that path

$$\oint E \cdot \mathrm{d}l = - \iint \frac{\partial B}{\partial t} \cdot \mathrm{d}A. \tag{1.4}$$

These are the integral forms of the equations. If the notation is found intimidating it is helpful to remember that it is just a way of writing the usual statements of the laws of electromagnetism in the shorthand notation of mathematics. The laws may also be written in equivalent, differential, forms:

1. Gauss' theorem of electrostatics

$$\nabla \cdot D = \varrho; \tag{1.5}$$

2. Gauss' theorem of magnetostatics

$$\nabla \cdot B = 0; \tag{1.6}$$

3. The magnetic circuit law

$$\nabla \wedge H = J + \frac{\partial D}{\partial t}; \text{ and} \tag{1.7}$$

4. Faraday's law of electromagnetic induction

$$\nabla \wedge E = -\frac{\partial B}{\partial t}. \tag{1.8}$$

The notation of vector calculus used above may, again, be rather intimidating to those who are not mathematically minded. The important thing to remember is that these expressions can be given meanings in a variety of systems of coordinates. Appendix B summarizes these interpretations in Cartesian, and cylindrical and spherical polar coordinates, those being the ones most commonly used by engineers. For the greater part of this book only rectangular Cartesian coordinates are required.

Maxwell's equations are believed to be expressions of basic physical laws. In order to make use of them we also require another set of equations which summarize experimental information about the properties of materials (Carter, 1986 p. 130).

1. Ohm's law

$$J = \sigma E, \tag{1.9}$$

where σ is the conductivity of the material. In this book σ is used rather than its reciprocal, the resistivity ϱ, to avoid confusion with the use of that symbol for charge density.
2. Dielectric materials

$$D = \varepsilon E, \tag{1.10}$$

where ε is the permittivity of the material which can also be written as

$$\varepsilon = \varepsilon_0 \varepsilon_r, \tag{1.11}$$

where ε_0 is the primary electric constant and ε_r is the relative permittivity of the material.
3. Magnetic materials

$$B = \mu H, \tag{1.12}$$

where μ is the permeability of the material which can also be written as

$$\mu = \mu_0 \mu_r, \tag{1.13}$$

where μ_0 is the primary magnetic constant and μ_r is the relative permeability of the material. For some materials either or both of ε_r and μ_r may be complex indicating a phase difference between D and E or B and H for alternating fields.

It is important to bear in mind that these equations are useful approximations of experimental results. They assume that the material properties are constants. While this is a satisfactory assumption for many conducting and dielectric materials it is only a crude approximation for ferromagnetic and ferroelectric materials. In some cases the material properties cannot be regarded as scalar quantities. The vectors which are related to each other by an equation are then not parallel to each other. These aspects of the subject are beyond the scope of this book. Here we shall be assuming that all the properties of materials can be described by scalar constants. (See Dekker, 1959, for further information)

The final equation which will be needed summarizes the law of conservation of charge:

The net current flow out of a closed surface is equal to the rate of decrease of the enclosed charge.

In mathematics this is expressed by the continuity equation which can be written in both integral and differential forms

$$\oiint \boldsymbol{J} \cdot d\boldsymbol{A} = -\frac{\partial}{\partial t} \iiint \varrho \, dv \tag{1.14}$$

$$\nabla \cdot \boldsymbol{J} = -\frac{\partial \varrho}{\partial t}. \tag{1.15}$$

The remainder of this chapter explores the possibility of plane wave solutions to Maxwell's equations in a variety of media.

1.3 ELECTROMAGNETIC WAVES IN NON-CONDUCTING MEDIA

The field theory of a lossless two-wire transmission line assumes that the electric and magnetic fields are perpendicular to each other and to the direction of the line (Carter, 1986, Ch. 7). The results of this theory are consistent with experiment and with the parallel approach using circuit theory summarized in Appendix A. A wave of this kind is called a transverse electric and magnetic (TEM) wave. It is natural to enquire whether this result can be generalized to other situations.

The exploration starts from equations (1.7) and (1.8). By restricting attention to non-conducting materials (1.7) can be simplified to give

$$\nabla \wedge \boldsymbol{H} = \frac{\partial \boldsymbol{D}}{\partial t}. \tag{1.16}$$

To simplify matters still further we assume that, working in rectangular Cartesian coordinates, $\boldsymbol{E}$ only has a component in the x direction and $\boldsymbol{H}$ only in the y direction. If we also assume that any wave propagates in the z direction it follows that the two field vectors vary only with z and t.

In rectangular Cartesian coordinates, the curl of $\boldsymbol{H}$ can be written in the form of a determinant (see Appendix B)

$$\nabla \wedge \boldsymbol{H} = \begin{vmatrix} \hat{x} & \hat{y} & \hat{z} \\ \partial/\partial x & \partial/\partial y & \partial/\partial z \\ H_x & H_y & H_z \end{vmatrix}, \tag{1.17}$$

where $\hat{x}$, $\hat{y}$ and $\hat{z}$ are unit vectors in the x, y and z directions, respectively. Evaluation of the determinant with the assumed direction of $\boldsymbol{H}$ yields

$$\nabla \wedge \boldsymbol{H} = -\hat{x} \, \partial H_y/\partial z \tag{1.18}$$

since $\partial H_y/\partial x$ and $\partial H_y/\partial y$ are both zero. Notice that this vector is in the x direction. Equation (1.16) therefore becomes

$$\frac{\partial H_y}{\partial z} = -\varepsilon \frac{\partial E_x}{\partial t}. \tag{1.19}$$

This is a scalar equation because the vectors on either side are both in the x direction.

In the same way equation (1.8) leads to

$$\frac{\partial E_x}{\partial z} = -\mu \frac{\partial H_y}{\partial t}. \tag{1.20}$$

These equations are remarkably similar in form to the telegrapher's equations derived from the circuit theory of transmission lines (Carter, 1986, p. 108)

$$\frac{\partial I}{\partial x} = -C \frac{\partial V}{\partial t} \tag{1.21}$$

$$\frac{\partial V}{\partial x} = -L \frac{\partial I}{\partial t}. \tag{1.22}$$

The resemblance is even more striking when the dimensions of the quantities are recalled: H in amps per metre, E in volts per metre, ε in farads per metre and μ in henries per metre.

Differentiating (1.19) with respect to t and (1.20) with respect to z gives

$$\frac{\partial^2 H_y}{\partial z \partial t} = -\varepsilon \frac{\partial^2 E_x}{\partial t^2} \tag{1.23}$$

$$\frac{\partial^2 E_x}{\partial z^2} = -\mu \frac{\partial^2 H_y}{\partial z \partial t}, \tag{1.24}$$

whence

$$\frac{\partial^2 E_x}{\partial z^2} = \varepsilon\mu \frac{\partial^2 E_x}{\partial t^2}. \tag{1.25}$$

This is the standard form of the wave equation. For sinusoidal waves the solution can be written (Carter, 1986, p. 110)

$$E_x = E_+ \exp j(\omega t - kz) + E_- \exp j(\omega t + kz), \tag{1.26}$$

where E_+ and E_- are the amplitudes of waves travelling in the positive and negative z directions, respectively. It should be remembered that the use of complex notation is a convenient way of simplifying the mathematical manipulations and that here, as in the 'j notation' used in circuit theory, it is the real part of every expression which is taken to have physical significance.

The phase velocity of the waves is given by

$$v_p = \frac{\omega}{k} = \frac{1}{\sqrt{(\varepsilon\mu)}} \tag{1.27}$$

(c.f. the relationship $v_p = 1/\sqrt{(LC)}$ for a transmission line).

For the special case of waves travelling in free space, the numerical value of the phase velocity calculated from experimental values of ε_0 and μ_0 agrees with the experimental measurements of the velocity of light within the limits of experimental error. It is easy to show that H_y also satisfies an equation like (1.25).

The relationship between E_x and H_y can be found by substituting the general solution (1.26) and the equivalent expression for H_y into equation (1.19). Then for waves travelling in the positive direction

$$jkH_y = j\omega\varepsilon E_x \qquad (1.28)$$

or

$$\frac{E_x}{H_y} = \sqrt{\left(\frac{\mu}{\varepsilon}\right)} = Z_w. \qquad (1.29)$$

This quantity has the dimensions of impedance and it is referred to as the wave impedance. In free space it has the numerical value $377\,\Omega$. One important deduction can be made from (1.29) namely that E_x and H_y are in phase with each other. It can be shown that the wave impedance of the wave in a transmission line also satisfies (1.29). (Carter, 1986). For waves travelling in the $-z$ direction k is negative and a minus sign appears in (1.28). The relationship between E_x and H_y in a plane electromagnetic wave is usually represented by the diagram shown in Fig. 1.2. This diagram shows the field vectors varying sinusoidally in space with the whole pattern moving in the z direction at the phase velocity of the wave. This diagram is a little misleading because it does not give the impression that E_y and H_y are constant over any plane perpendicular to the z axis. At this point it may be asked how it is possible for the field lines to lie exactly in a plane and,

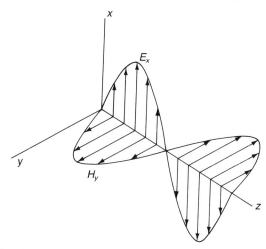

Fig. 1.2 Relationship between the electric and magnetic field vectors in a plane electromagnetic wave.

consequently, never end. The answer to this is that, in practice, a plane wave is created by launching a spherical wave. At a large distance from the source any small part of the wave front is effectively a plane wave.

1.4 ENERGY FLOW IN AN ELECTROMAGNETIC WAVE

When a wave $V \exp \mathrm{j}(\omega t - kz)$ propagates on a transmission line of characteristic impedance Z_0 the instantaneous flow of power along the line given by

$$P = VI = V^2/Z_0 = Z_0 I^2. \tag{1.30}$$

It is plausible to suppose that, in the same way, an electromagnetic wave propagating in free space also carries power. Now in a region where the electric and magnetic field strengths are E and H the stored energy density is

$$W = \tfrac{1}{2}\varepsilon|E|^2 + \tfrac{1}{2}\mu|H|^2. \tag{1.31}$$

By making use of the relationship (1.29) between E and H we find that the peak energy density in an electromagnetic wave is

$$W = \sqrt{(\varepsilon\mu)}\,|E|\,|H|. \tag{1.32}$$

The peak power flow is obtained by multiplying this expression by the velocity of the wave from (1.27) to give

$$S = |E|\,|H|. \tag{1.33}$$

Strictly speaking the group velocity should be used here in place of the phase velocity (see p. 40) but for waves propagating in uniform dielectric media they are identical.

Because the power flow is a vector which is perpendicular to both E and H it is useful to write (1.33) in vector form

$$S = E \wedge H. \tag{1.34}$$

The vector S is known as Poynting's vector. Since E, H and S are in the x, y and z directions respectively equation (1.34) gives the correct direction for S.

When S is integrated over a closed surface the principle of the conservation of energy requires that the total power flow should be equal to the rate of change of energy stored. This statement is known as Poynting's theorem. The proof of the theorem involves advanced mathematical techniques so it is not given here; it can be found in standard texts (e.g. Ramo *et al.*, 1965). It should be noted that the proof only gives a physical significance to the flux of S out of a closed surface and not to S itself. Where energy is dissipated as heat within the surface that must also be taken into account in applying the principle of conservation of energy. Poynting's theorem can be regarded as a generalization of the ideas of the flow of energy in electric circuits which associates the flow with the fields rather than directly with the currents and voltages.

The Poynting vector is not the only possible expression of the power flow in an electromagnetic wave. An alternative, the Slepian vector, is discussed by Carter (1967). The two approaches correspond to rather different physical pictures of the way in which energy is transmitted and dissipated. In practice the Poynting vector is the one generally used.

Very often the electromagnetic power flow is that in a sinusoidal electromagnetic wave. In that case the time average Poynting vector is

$$\langle S \rangle = \tfrac{1}{2} E \wedge H^* \tag{1.35}$$

where E and H are the complex wave vectors, the asterisk indicates the complex conjugate, and the factor of $\tfrac{1}{2}$ is a consequence of averaging the power flow over a full cycle of the wave. This expression is closely analogous to the usual expression for the power flow in an electric circuit

$$P = \tfrac{1}{2} V I^* \tag{1.36}$$

demonstrating again that the equations of circuit theory are special cases of the general laws of electromagnetism which apply when the currents are constrained to flow in wires and the components of the circuit can be regarded as lumped.

1.5 ELECTROMAGNETIC WAVES IN CONDUCTING MATERIALS

If the wave propagates in a conducting material then it is necessary to include the conduction current density J in the equations. If we again assume that the electric and magnetic fields are in the x and y directions respectively and that they vary as $\exp j(\omega t - kz)$ then (1.7) becomes

$$jkH_y = (\sigma + j\omega\varepsilon)E_x. \tag{1.37}$$

A good conductor may be defined as a material in which the conduction current is much greater than the displacement current, that is $\sigma \gg j\omega\varepsilon$ in (1.37) so that, approximately

$$jkH_y = \sigma E_x. \tag{1.38}$$

For copper $\sigma = 5.7 \times 10^7 \, \mathrm{S\,m^{-1}}$ and $\varepsilon = 10^{-11} \, \mathrm{F\,m^{-1}}$ so that the approximation is valid up to frequencies around $10^{16} \, \mathrm{Hz}$. Similarly, from (1.8) we get

$$jkE_x = j\omega\mu H_y \tag{1.39}$$

so that

$$k^2 = -j\omega\sigma\mu$$
$$k = \pm\sqrt{-j}\,\sqrt{(\omega\sigma\mu)}. \tag{1.40}$$

The square root of $-j$ can be found by noting that

$$-j = \exp(-j\pi/2) \tag{1.41}$$

so that

$$
\begin{aligned}
\sqrt{-j} &= \exp(-j\pi/4) \\
&= \cos(-\pi/4) + j\sin(-\pi/4) \\
&= \frac{1}{\sqrt{2}}(1 - j)
\end{aligned}
\tag{1.42}
$$

and

$$
\begin{aligned}
k &= \pm(1 - j)\sqrt{\left(\frac{\omega\sigma\mu}{2}\right)} \\
&= \pm(1 - j)/\delta,
\end{aligned}
\tag{1.43}
$$

where

$$\delta = \sqrt{(2/\omega\sigma\mu)}. \tag{1.44}$$

Substituting for k in the expressions for the fields shows that E and H vary (for waves travelling in the $+z$ direction) as

$$\exp j(\omega t - z/\delta)\,\exp(-z/\delta).$$

Thus a wave propagates with an exponentially decreasing amplitude. The decay constant δ is known as the skin depth for reasons which will become apparent in Chapter 4. The skin depth varies with frequency and with the properties of the material. For copper we find

$$
\begin{aligned}
&\text{at } 1\,\text{kHz} &&\delta = 2\,\text{mm} \\
&\text{at } 1\,\text{MHz} &&\delta = 67\,\mu\text{m} \\
&\text{at } 1\,\text{GHz} &&\delta = 2\,\mu\text{m}.
\end{aligned}
$$

The wave impedance is given by

$$
\begin{aligned}
Z_w &= E_x/H_y = jk/\sigma \\
&= (1 + j)/\sigma\delta
\end{aligned}
\tag{1.45}
$$

so that the electric and magnetic fields are not in phase with each other but the electric field leads the magnetic field by $45°$.

The power flow in the wave is given by the real part of the complex Poynting vector

$$
\begin{aligned}
|S| &= \text{Re}|\tfrac{1}{2}E \wedge H^*| \\
&= \text{Re}\left[\tfrac{1}{2}|E|^2\left(\frac{\sigma\delta}{1 + j}\right)\right] \\
&= \text{Re}[\tfrac{1}{4}|E|^2\sigma\delta(1 - j)] \\
&= \tfrac{1}{4}|E|^2\sigma\delta.
\end{aligned}
\tag{1.46}
$$

In some materials, mainly lossy dielectrics, the conduction and displacement currents are of comparable magnitudes so that the solution for k must be derived from (1.37) and (1.39). An added complication is that, because of the effects of vibration and rotation of molecules, D and E are not necessarily in phase with each other. This is allowed for by writing

$$\varepsilon = \varepsilon' - j\varepsilon''$$

so that (1.37) becomes

$$\begin{aligned} jkH_y &= [(\sigma + \omega\varepsilon'') + j\omega\varepsilon']E_x \\ &= (\sigma' + j\omega\varepsilon')E_x. \end{aligned} \qquad (1.47)$$

The relationship between the conduction and displacement currents is illustrated in Fig. 1.3. The angle between the total current and the displacement current is given by

$$\tan \delta = \sigma'/\omega\varepsilon'. \qquad (1.48)$$

This quantity is known as the loss tangent of the material. Note carefully that δ here is not the skin depth referred to above. Equation (1.47) can therefore be written

$$jkH_y = (1 - j \tan \delta)j\omega\varepsilon'E_x \qquad (1.49)$$

so that

$$\begin{aligned} k^2 &= (1 - j \tan \delta)\omega^2\varepsilon'\mu \\ k &= \sqrt{(1 - j \tan \delta)}\omega\sqrt{(\varepsilon'\mu)} \end{aligned} \qquad (1.50)$$

for many materials the loss tangent is small so that

$$k \simeq (1 - \tfrac{1}{2}j \tan \delta)\omega\sqrt{(\varepsilon'\mu)}. \qquad (1.51)$$

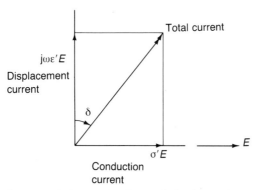

Fig. 1.3 Phasor diagram of the conduction and displacement currents in a lossy dielectric material.

Substituting this expression into the solution assumed for the propagation of the waves we find

$$\exp j\omega[t - \sqrt{(\varepsilon'\mu)}z] \exp -[\tfrac{1}{2}\omega \cdot \sqrt{(\varepsilon'\mu)} \tan \delta]z \qquad (1.52)$$

so that the waves decay as they propagate through the material. Typical materials include plastics, ceramics and many organic materials. This part of the theory finds application in the use of electromagnetic energy in microwave ovens. For example, steak has $\varepsilon' = 40\varepsilon_0$ and $\tan \delta = 0.3$ at 3 GHz. Substituting these figures into the expression above shows that the fields fall off by a factor $1/e$ in 17 mm.

1.6 PROPAGATION OF WAVES IN PLASMAS

So far it has been assumed that the conduction charges in a conductor are able to respond instantly to the field of an electromagnetic wave. This assumption is not always valid. A particular case is that of an ionized gas in which there are two species of mobile charge carriers, free electrons and the very much more massive positive ions. It is convenient to speak of such a gas as a plasma although some would restrict the use of that term to situations where the ions are completely stripped of their electrons. The theory of the interaction between electromagnetic waves and plasmas is important for understanding the propagation of waves in the ionosphere, in electron devices and in experiments in thermonuclear fusion. Some of these cases are rather difficult because they involve the random thermal motions of the charge carriers and collisions between them. Here the basic ideas are illustrated by considering a cold, collisionless plasma.

We assume, as before, that the wave is a pure TEM wave propagating in the z direction so that

$$E_x = E_0 \exp j(\omega t - kz)$$

and
$$H_y = H_0 \exp j(\omega t - kz). \qquad (1.53)$$

To make things simpler we will also assume that the ions can be regarded as fixed so that only the electron motion has to be considered. The current density, charge density and electron velocity at a point in the plasma are related by

$$J = \varrho v. \qquad (1.54)$$

The plasma as a whole is supposed to be electrically neutral so only the time-varying part of this equation matters, that is

$$J_1 = \varrho_0 v_1 + \varrho_1 v_1. \qquad (1.55)$$

Provided that the signal levels are small $\varrho_1 \ll \varrho_0$ and the second term is negligible compared with the first so that

$$J_1 = \varrho_0 v_1. \tag{1.56}$$

It can be shown that the magnetic forces on the electrons are much smaller than the electric forces so that, effectively,

$$m\ddot{x} = qE_x$$

or

$$j\omega m v_1 = qE_x. \tag{1.57}$$

Eliminating v_1 between (1.56) and (1.57) gives

$$J_1 = -j\frac{\eta\varrho_0}{\omega}E_x, \tag{1.58}$$

where $\eta = q/m$ is the charge to mass ratio of the electron. When the current and the magnetic field from (1.53) are substituted into (1.7) we get

$$jkH_y = j\omega\varepsilon_0\left(1 - \frac{\omega_p^2}{\omega^2}\right)E_x, \tag{1.59}$$

where

$$\omega_p^2 = \eta\varrho_0/\varepsilon_0. \tag{1.60}$$

This frequency is known as the plasma frequency.

A second relationship between the two field vectors is given by (1.39) and they may then be eliminated to give the propagation constant of the wave

$$k = \omega\sqrt{(\varepsilon_0\mu_0)}\left(1 - \frac{\omega_p^2}{\omega^2}\right)^{\frac{1}{2}}. \tag{1.61}$$

From this it is clear that the propagation constant has real values only when the signal frequency is greater than the plasma frequency. The plasma then behaves as a dielectric medium whose permittivity, from comparison with (1.27), is

$$\varepsilon = \varepsilon_0\left(1 - \frac{\omega_p^2}{\omega^2}\right). \tag{1.62}$$

At frequencies below the plasma frequency k is pure imaginary so that the waves decay as

$$\exp\left[-\left(\frac{\omega_p^2}{\omega^2} - 1\right)^{\frac{1}{2}}k_0z\right], \tag{1.63}$$

where k_0 is the free-space propagation constant at frequency ω. Further light is shed on this behaviour by consideration of the wave impedance and of the relationship between the conduction and displacement current densities. The wave impedance is, from (1.59) and (1.61),

$$Z_w = \sqrt{\left(\frac{\mu_0}{\varepsilon_0}\right)}\left(1 - \frac{\omega_p^2}{\omega^2}\right)^{\frac{1}{2}}. \tag{1.64}$$

Above the plasma frequency this is real so that E and H are in phase with each other. Below the plasma frequency it is pure imaginary so that E and H are in phase quadrature and there is no net flow of power. The displacement current is

$$J_d = \frac{\partial D}{\partial t} = j\omega\varepsilon_0 E_x, \tag{1.65}$$

so that, from (1.58), the conduction current is

$$J_1 = -\frac{\omega_p^2}{\omega^2} J_d. \tag{1.66}$$

Thus below the plasma frequency the electrons can follow the wave, the conduction current dominates and the plasma behaves as a conductor. Above the plasma frequency they cannot do so, the diplacement current dominates and the plasma behaves as a dielectric.

Similar behaviour is seen in some dielectric materials which contain molecules having an electric dipole moment. The permittivities of these materials vary with frequency according to whether the molecules can rotate to follow the changing electric field or not. For further information consult the book by Bleaney and Bleaney (1976).

1.7 POLARIZATION OF WAVES

So far we have only discussed waves which have their electric and magnetic field vectors in the x and y directions, respectively. It is obvious that this is not the only possible orientation of the field vectors for a wave propagating in the positive z direction. The orientation of the electric field vector is referred to by electronic engineers as the plane of polarization of the wave. Somewhat confusingly the convention adopted in optics is to define the plane of polarization as the direction of the magnetic field vector. In this book the first convention will be used throughout.

Any general direction of polarization can be considered as a superposition of two waves having the same phase as each other and polarized in the x and y directions as shown in Fig. 1.4. Such a wave is known as a plane-polarized wave. We shall see later that waves having different polarizations behave differently when they pass through certain media and when they are reflected from the interface between two dielectric materials. These properties have important practical consequences.

Now consider the slightly more complicated case where the two components of a wave are equal in amplitude but have phases which differ from each other by 90°. At a particular plane perpendicular to the z axis the electric field of the wave is

$$E = \hat{x}E_0 \cos \omega t + \hat{y}E_0 \sin \omega t \tag{1.67}$$

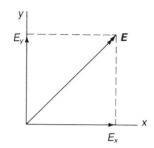

Fig. 1.4 Combination of waves polarized in the x and y directions.

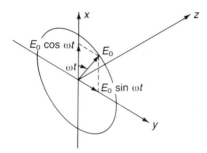

Fig. 1.5 Relationship between the x and y components of the electric field in a circularly polarized wave.

so that the tip of the electric field vector is rotating around a circle with angular velocity ω as shown in Fig. 1.5. The wave is then said to be circularly polarized. Moreover, because the direction of rotation of the electric field vector with time is in the right hand corkscrew sense with respect to the z axis, it is positive circularly polarized. Evidently, if the phase difference between the x and y components had been made $-90°$ the sense of rotation would have been reversed and the resulting wave would have been negative circularly polarized.

In the most general case of all the amplitudes of the two components may differ from each other and their phases be other than in quadrature. The tip of the electric field vector then traces out an ellipse and the wave is said to be elliptically polarized. As this increase in generality introduces no new principles it will not be pursued further here and the reader is referred to more advanced texts for the details (e.g. Jordan and Balmain, 1968; Longhurst, 1973)

1.8 PROPAGATION IN GYROMAGNETIC MEDIA

Some media have the property that when they are placed in a steady magnetic field the propagation constants for positive and negative circularly polarized waves differ from each other. Examples are ionized gasses and ferrites. They are known collectively as gyromagnetic materials for reasons which will become apparent. Ferrites are made by sintering mixtures of oxides of iron and of metals such as nickel or magnanese (Baden-Fuller, 1987). They combine ferromagnetic properties with high electrical resistivity and have been developed because of their usefulness at high frequencies. The mathematical treatment of wave propagation in ferrites is rather involved so the case of propagation in an ionized gas in a magnetic field will be used here to illustrate how the gyromagnetic properties arise. Wave propagation in ferrites is discussed in Chapter 8.

Consider, then, an ionized gas which is in a steady magnetic field B_0 directed parallel to the z axis. Let a positive circular polarized wave pass through in the z direction. The electric field of the wave is then

$$E = [\hat{x} - j\hat{y}]E_0 \exp j(\omega t - kz). \qquad (1.68)$$

To simplify the derivation we assume that each electron moves in a circular orbit perpendicular to the z axis with an angular velocity ω. The ions are assumed to be sufficiently massive so that their motion can be ignored. The electrons experience the rotating radial electric field of the wave shown in Fig. 1.5 and the condition for a steady orbit is that the radial forces should balance. That is

$$-qE_0 - qr\omega B_0 + mr\omega^2 = 0, \qquad (1.69)$$

where q is the magnitude of the charge on an electron. The three terms represent the electric force, the magnetic force produced by the motion of the electron through the magnetic field, and the centrifugal force. Equation (1.69) can be rearranged to give the radius of the stable orbit

$$r = \frac{\eta E_0}{\omega(\omega - \omega_c)}, \qquad (1.70)$$

where η is the charge to mass ratio of the electron and

$$\omega_c = \eta B_0 \qquad (1.71)$$

is the cyclotron frequency, that is, the angular velocity of the electron for a stable orbit when the electric field is zero. Equation (1.70) shows that the radius of the orbit is greatest when the signal frequency is equal to the cyclotron frequency. It would not become infinite in practice because of the effects of collisions. When $\omega > \omega_c$ r is positive so the electron moves in phase with the applied field. When $\omega < \omega_c$ r is negative, implying that the motion and the field are in antiphase.

Figure 1.6 shows the Cartesian components of the velocity of the electron. If there are N electrons per unit volume then the x component of the current density is

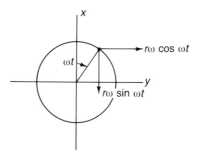

Fig. 1.6 Cartesian components of the velocity of an electron moving in a circular orbit.

$$J_x = -Nq\dot{x}$$
$$= -jNq\omega r \exp j(\omega t - kz). \tag{1.72}$$

Then, from (1.7), we get

$$jkH_y = J_x + j\omega\varepsilon_0 E_x \tag{1.73}$$

or

$$kH_y = \omega\varepsilon_0 E_x \left[1 - \frac{\omega_p^2}{\omega(\omega - \omega_c)} \right]. \tag{1.74}$$

Combining this with (1.39) gives the propagation constant of the wave (c.f. eqn (1.61))

$$k_+ = \omega\sqrt{(\varepsilon_0\mu_0)} \left[1 - \frac{\omega_p^2}{\omega(\omega - \omega_c)} \right]^{\frac{1}{2}}, \tag{1.75}$$

where the subscript + indicates positive circular polarization. Equation (1.75) shows that at very high frequencies the propagation constant tends to that of free space. This is because the electrons are unable to follow the changes in the field.

The equivalent expression for the negative circularly polarized wave is obtained by setting ω equal to $-\omega$, giving

$$k_- = \omega\sqrt{(\varepsilon_0\mu_0)} \left[1 - \frac{\omega_p^2}{\omega(\omega + \omega_c)} \right]^{\frac{1}{2}}. \tag{1.76}$$

Thus the positively and negatively polarized waves have different propagation constants. The physical explanation for this effect is revealed when equation (1.69) is examined. For positive rotation the magnetic force is inwards whilst for negative rotation it is outwards. Thus for positive rotation the electric force required to produce equilibrium can be either inwards or outwards, whereas for negative rotation it must always be inwards.

This effect has important practical consequences. A plane polarized wave can be considered as the superposition of a pair of circularly polarized waves by writing

$$\hat{x}E_x = \tfrac{1}{2}(\hat{x} + j\hat{y})E_x + \tfrac{1}{2}(\hat{x} - j\hat{y})E_x, \qquad (1.77)$$

where the two brackets on the right hand side of the equation represent a pair of positive and negative circularly polarized waves.

In most media the propagation constants of the two waves are the same and at any other plane they can be recombined to give a plane polarized wave with the same plane of polarization as before. In a gyromagnetic medium, however, the propagation constants of the waves differ, as we have seen, so that the phase relationship between them changes as they propagate. When they are recombined the effect is to produce a plane polarized wave whose plane of polarization has been rotated about the z axis relative to the initial polarization. This effect is known as Faraday rotation.

In the ionosphere the Earth's atmosphere is ionized by cosmic rays to produce a plasma which is influenced by the Earth's magnetic field. Thus the ionosphere is a gyromagnetic medium of the kind discussed above. Radio signals transmitted from satellites experience a rotation in their planes of polarization as they pass through the ionosphere and the extent of the rotation varies with time because of variations in the density of free electrons. If plane waves were used for the transmissions there would be difficulties in receiving them because of this effect. The solution is to transmit circularly polarized waves so that the Faraday rotation appears only as a phase shift in the signal received.

Faraday rotation also occurs when a wave passes through a ferrite material which is in a steady magnetic field. This is put to use in the microwave devices known as circulators and isolators which are discussed further in Chapter 8.

1.9 BOUNDARY CONDITIONS

The study of waves in uniform, infinite media is of limited interest. Practical problems usually involve more than one medium so that the behaviour of the waves at the interfaces is very important. The starting point for discussing this subject is the statement of the boundary conditions which apply to the electric and magnetic fields at a boundary at which there are no surface charges or currents.

1. The tangential component of the electric field E is continuous.
2. The normal component of the electric flux density D is continuous.
3. The tangential conponent of the magnetic field H is continuous.
4. The normal component of the magnetic flux density B is continuous.

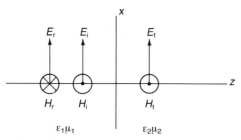

Fig. 1.7 Incident, reflected and transmitted wave fields for an electromagnetic wave incident normally on a dielectric boundary.

The proofs of these conditions can be found in books on elementary electromagnetism (Carter, 1986, pp. 24 and 66).

The simplest case of interaction between an electromagnetic wave and a boundary arises when the direction of propagation is normal to it. The field vectors are then parallel to the boundary and only boundary conditions 1 and 3 are needed. Figure 1.7 shows this situation with the incident, transmitted and reflected wave fields. The origin of coordinates is taken to lie on the boundary for convenience. The propagation constants in the two materials are

$$k_1 = \omega\sqrt{(\varepsilon_1\mu_1)} \quad \text{and} \quad k_2 = \omega\sqrt{(\varepsilon_2\mu_2)} \tag{1.78}$$

and the wave impedances are

$$Z_1 = \sqrt{\left(\frac{\mu_1}{\varepsilon_1}\right)} \quad \text{and} \quad Z_2 = \sqrt{\left(\frac{\mu_2}{\varepsilon_2}\right)}. \tag{1.79}$$

The three waves are then

$$E_i \exp j(\omega t - k_1 z)$$
$$E_r \exp j(\omega t + k_1 z)$$
$$E_t \exp j(\omega t - k_2 z). \tag{1.80}$$

Boundary condition 1 requires that the electric fields should be the same on both sides of the boundary.

$$E_i + E_r = E_t. \tag{1.81}$$

Similarly condition 3 yields

$$H_i - H_r = H_t. \tag{1.82}$$

The equations are analogous to those for the voltage and current at a discontinuity in a transmission line (Appendix A). The magnetic field vectors for the incident and reflected waves must be of opposite sign in order to give the correct directions for the power flow. By making use of the wave impedances (1.82) can be rewritten

$$(E_i - E_r)/Z_1 = E_t/Z_2, \qquad (1.83)$$

whence the reflected wave is given by

$$\frac{E_r}{E_i} = \frac{Z_2 - Z_1}{Z_2 + Z_1} \qquad (1.84)$$

and the transmitted wave is given by

$$\frac{E_t}{E_i} = \frac{2Z_2}{Z_2 + Z_1}. \qquad (1.85)$$

We have seen in the preceding sections that the wave impedance is sometimes a complex or imaginary quantity. When this occurs the ratios of the wave amplitudes are complex indicating reflection and transmission with a change of phase. An important special case arises when the second material is regarded as a perfect conductor. The electric field within it and the wave impedance must then be zero so that

$$\frac{E_r}{E_i} = -1 \quad \text{and} \quad \frac{E_t}{E_i} = 0 \qquad (1.86a)$$

that is, the wave is totally reflected at the boundary with the reflected wave in antiphase with the incident wave. The corresponding equations for the magnetic fields are

$$\frac{H_r}{H_i} = 1 \quad \text{and} \quad \frac{H_t}{H_i} = 0. \qquad (1.86b)$$

From these it follows that the magnetic field is $2H_i$ just outside the conductor and zero within it. There is, therefore, an apparent violation of boundary condition 3. The explanation is that currents flow in the surface of the conductor to match the fields on either side (see Carter, 1986, p. 66). The surface current density is equal to the tangential component of the magnetic field.

Equations (1.84) and (1.85) bear a striking resemblance to the equations for the transmission and reflection coefficients at a junction between two transmission lines of different impedances (see Carter, 1986, p. 112). This is not surprising when we remember that the waves on the lines can be described either in circuit terms or as TEM field waves. We have already seen that there is a close resemblance between the differential equations governing the two descriptions (eqns (1.19) to (1.21)). The analogy is very important practically because it allows us to make use of transmission line techniques for solving electromagnetic wave problems. This point will be explored further in Chapters 3 and 4.

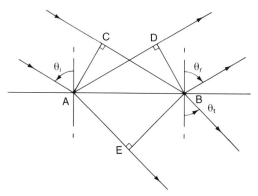

Fig. 1.8 Geometry of the reflection and refraction of waves at a dielectric boundary for oblique incidence.

In general a wave will not be incident normally on a boundary. For other angles of incidence we have to consider separately waves polarized normal to the boundary and waves polarized parallel to the boundary. Any more general case can be regarded as a superposition of these two.

First we shall establish the laws of reflection and refraction by considering Fig. 1.8. The diagram shows general incident, reflected and transmitted waves making angles θ_i, θ_r and θ_t to the normal to the boundary. Points A and B are chosen on the boundary so that the phase difference between them is 360° for the incident wave. Whatever conditions apply at A must also apply at B so that the phase differences for the other two waves must also be 360°. Wavefronts AC, BD and BE are constructed for each wave perpendicular to the directions of propagation. The distances CB, AD and AE are then the distances travelled by the wavefronts in one cycle of the wave, that is

$$CB = AD = \lambda_1$$

and
$$AE = \lambda_2, \tag{1.87}$$

where λ_1 and λ_2 are the wavelengths in the two media. The triangles ABC and ABD are similar triangles and the angles of incidence and reflection are proved to be equal to each other. We can therefore use subscripts 1 and 2 to refer to the angles in the two media.

For the transmitted wave, from triangles ABC and ABE we have

$$AB = \frac{\lambda_1}{\sin \theta_1} = \frac{\lambda_2}{\sin \theta_2} \tag{1.88}$$

or, since

$$\lambda_1 = \frac{2\pi}{\omega\sqrt{(\varepsilon_1\mu_1)}} \quad \text{and} \quad \lambda_2 = \frac{2\pi}{\omega\sqrt{(\varepsilon_2\mu_2)}} \tag{1.89}$$

$$\frac{\sin \theta_1}{\sin \theta_2} = \sqrt{\left(\frac{\varepsilon_2 \mu_2}{\varepsilon_1 \mu_1}\right)}. \tag{1.90}$$

This is evidently related to Snell's law of geometrical optics (Longhurst, 1973).

$$\frac{\sin \theta_1}{\sin \theta_2} = \frac{n_2}{n_1}, \tag{1.91}$$

where n_1 and n_2 are the refractive indexes of the two materials. The refractive index is the ratio of the velocity of light in free space to that in the medium. For non-ferromagnetic materials $\mu = \mu_0$ to a close approximation so that $n = \sqrt{\varepsilon_r}$.

Having established these basic relations we can proceed to consider the oblique incidence of a wave on a boundary taking first the case shown in Fig. 1.9 where the plane of polarization is normal to the boundary. Applying boundary conditions 1 and 3 as before yields

$$(E_i + E_r) \cos \theta_1 = E_t \cos \theta_2 \tag{1.92}$$

$$H_i - H_r = H_t. \tag{1.93}$$

Comparison with equations (1.81) and (1.82) shows that we can maintain the correspondence with transmission line theory if we define the normal wave impedance of the waves as

$$Z_{n\perp} = \frac{E \cos \theta}{H} = \sqrt{\left(\frac{\mu}{\varepsilon}\right)} \cos \theta. \tag{1.94}$$

When the electric field vectors are parallel to the boundary as shown in Fig. 1.10 the boundary conditions are

$$E_i + E_r = E_t \tag{1.95}$$

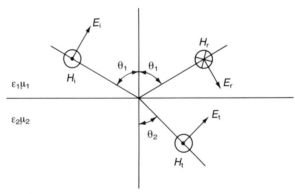

Fig. 1.9 Field vectors for oblique incidence of waves on a dielectric boundary with the electric field normal to the boundary.

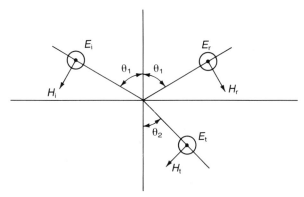

Fig. 1.10 Field vectors for oblique incidence of waves on a dielectric boundary with the electric field parallel to the boundary.

and
$$(H_i - H_r) \cos \theta_1 = H_t \cos \theta_2 \tag{1.96}$$

which can be made equivalent to (1.81) and (1.83) by defining the normal impedances of the waves by

$$Z_{n\,\|} = \frac{E}{H \cos \theta} = \sqrt{\left(\frac{\mu}{\varepsilon}\right)} \frac{1}{\cos \theta}. \tag{1.97}$$

Practical cases involving the interactions between waves and boundaries between different materials will be considered in later chapters.

1.10 CONCLUSION

In this chapter we have considered the propagation of plane electromagnetic waves through different media and seen how it depends upon their physical properties. The different cases considered are not exhaustive but have been chosen to illustrate the principal kinds of phenomena which occur. We have also established the laws of reflection and refraction which apply to waves at the interface between two materials. A close resemblance to transmission line theory has been demonstrated which promises to allow transmission line methods to be applied to problems involving electromagnetic waves. The fundamental concepts considered in this chapter are applied to practical situations in the chapters which follow.

EXERCISES

1.1 Calculate the wave impedances of electromagnetic waves travelling in free space, polystyrene ($\varepsilon_r = 2.7$), alumina ($\varepsilon_r = 8.9$) and Barium strontium titanate ($\varepsilon_r = 10\,000$).

1.2 Calculate the power density in an electromagnetic wave whose electric field strength is $100\,\mathrm{V\,m^{-1}}$ in the same materials as question 1.1.

1.3 Calculate the skin depth at $50\,\mathrm{Hz}$, $5\,\mathrm{MHz}$ and $5\,\mathrm{GHz}$ for silver ($\sigma = 6.1 \times 10^7\,\mathrm{S}$), Graphite ($\sigma = 10^5\,\mathrm{S}$) and seawater ($\sigma = 4\,\mathrm{S}$).

1.4 Calculate the attenuation in decibels per metre for electromagnetic waves at a frequency of $10\,\mathrm{GHz}$ travelling through glass ($\varepsilon_r = 4$, $\varepsilon''/\varepsilon' = 21 \times 10^{-4}$) and through fused quartz ($\varepsilon'/\varepsilon_0 = 3.78$, $\varepsilon''/\varepsilon' = 10^{-4}$).

1.5 Calculate the plasma frequency for electrons and for hydrogen molecular ions (mass = 3672 × mass of electron) when the particle densities are $10^{12}\,\mathrm{m^{-3}}$ and $10^{16}\,\mathrm{m^{-3}}$.

1.6 Calculate the wave impedances of the electron plasmas in question 1.5 at a frequency of $1\,\mathrm{GHz}$.

1.7 Calculate the electron cyclotron frequency at magnetic field strengths of 0.05, 0.1 and $0.2\,\mathrm{T}$.

1.8 Calculate the propagation constants at a frequency of $500\,\mathrm{MHz}$ for positive and negative circularly polarized waves in an electron plasma whose electron density is $10^{16}\,\mathrm{m^{-3}}$ in the presence of a magnetic field of $0.01\,\mathrm{T}$.

Waves guided by perfectly conducting boundaries | 2

2.1 TEM TRANSMISSION LINES

The simplest systems for guiding electromagnetic waves are the two-wire lines. Figure 2.1 illustrates some common types. These lines all have at least two conductors which are electrically insulated from each other. In addition the whole space where the electric field is not zero is filled with a uniform dielectric material. A number of other types of two-wire line exist which have more than one dielectric around them. These lines which have rather more complicated behaviour are discussed in the next chapter.

Elementary treatments of the theory of lines such as those shown in Fig. 2.1 assume that the fields are transverse. It is straightforward to show that this is consistent with the circuit approach for the special case of the coaxial cable (Carter, 1986). We now must prove that this result holds for all lines of this type.

The fields around the conductors obey Maxwell's equations (1.8) and (1.16). Taking the curl of (1.8) and using (1.16) gives

$$\nabla \wedge (\nabla \wedge E) = -\frac{\partial}{\partial t}(\nabla \wedge B)$$

$$= -\mu \frac{\partial}{\partial t}(\nabla \wedge H)$$

$$= -\varepsilon \mu \frac{\partial^2 E}{\partial t^2}. \qquad (2.1)$$

The left hand side of this equation can be written

$$\nabla \wedge (\nabla \wedge E) = \nabla(\nabla \cdot E) - \nabla^2 E. \qquad (2.2)$$

This relationship can be proved by evaluating the derivatives in terms of the Cartesian components of E though the task is a bit laborious. Now, since there is no free charge between the conductors $\varrho = 0$ and therefore $\nabla \cdot E = 0$ from (1.5). Thus from (2.1) and (2.2) we have

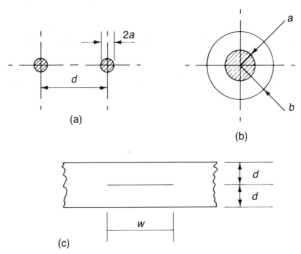

Fig. 2.1 Two-wire transmission lines: (a) parallel-wire line, (b) coaxial cable, and (c) triplate line.

$$\nabla^2 \mathbf{E} = \varepsilon\mu \frac{\partial^2 \mathbf{E}}{\partial t^2}. \tag{2.3}$$

This is the three-dimensional form of the wave equation. In Cartesian coordinates it can be written

$$\frac{\partial^2 E_x}{\partial x^2} + \frac{\partial^2 E_x}{\partial y^2} + \frac{\partial^2 E_x}{\partial z^2} = \varepsilon\mu \frac{\partial^2 E_x}{\partial t^2} \tag{2.4}$$

together with similar equations for E_y and E_z. Now suppose that $\mathbf{E}$ has only x and y components and that these vary as $\exp j(\omega t - kz)$. Substitution into (2.4) produces

$$\frac{\partial^2 E_x}{\partial x^2} + \frac{\partial^2 E_x}{\partial y^2} - k^2 E_x = -\varepsilon\mu\omega^2 E_x. \tag{2.5}$$

If we assume that

$$k^2 = \varepsilon\mu\omega^2. \tag{2.6}$$

Then

$$\frac{\partial^2 E_x}{\partial x^2} + \frac{\partial^2 E_x}{\partial y^2} = 0. \tag{2.7}$$

But $\mathbf{E}$ can be written in terms of a scalar potential V

$$E_x = -\frac{\partial V}{\partial x} \tag{2.8}$$

so that

$$\frac{\partial^2}{\partial x^2}\left(\frac{\partial V}{\partial x}\right) + \frac{\partial^2}{\partial y^2}\left(\frac{\partial V}{\partial x}\right) = 0$$

$$\frac{\partial}{\partial x}\left(\frac{\partial^2 V}{\partial x^2} + \frac{\partial^2 V}{\partial y^2}\right) = 0. \tag{2.9}$$

Similarly by considering the y component of E we obtain

$$\frac{\partial}{\partial y}\left(\frac{\partial^2 V}{\partial x^2} + \frac{\partial^2 V}{\partial y^2}\right) = 0. \tag{2.10}$$

Equations (2.9) and (2.10) can be satisfied simultaneously if the expression in the brackets is zero, that is, if V satisfies Laplace's equation (Carter, 1986, p. 16). Thus for the mode of propagation which satisfies (2.6) the electric field distribution is identical to the electrostatic field between the electrodes. That is why it is possible to compute the capacitance and inductance per unit length of such a line from the static field solutions. It should be noted that, whilst this TEM wave is a possible solution, there could be other solutions for which (2.6) is not satisfied and the fields differ from the static fields. This point will be examined further in Section 2.8.

The use of field theory here, as very often in electromagnetism, is to provide a way of calculating the circuit parameters. The phase velocity of a TEM wave on a two-wire line is constant and equal to that of an unbounded plane TEM wave propagating through the same medium as that which separates the conductors. Expressions for the capacitance and inductance per unit length and the characteristic impedance for the two-wire lines shown in Fig. 2.1 are given in Table 2.1.

Table 2.1

Line	C (Fm^{-1})	L (Hm^{-1})	Z (Ω)
Coaxial	$\dfrac{2\pi\varepsilon}{\ln(b/a)}$	$\dfrac{\mu}{2\pi}\ln\left(\dfrac{b}{a}\right)$	$\dfrac{1}{2\pi}\sqrt{\left(\dfrac{\mu}{\varepsilon}\right)}\ln\left(\dfrac{b}{a}\right)$
Two wire	$\dfrac{\pi\varepsilon}{\cosh^{-1}\left(\dfrac{d}{2a}\right)}$	$\dfrac{\mu}{\pi}\cosh^{-1}\left(\dfrac{d}{2a}\right)$	$\dfrac{1}{\pi}\sqrt{\left(\dfrac{\mu}{\varepsilon}\right)}\cosh^{-1}\left(\dfrac{d}{2a}\right)$
Triplate $w \gg d$	$\dfrac{2\varepsilon w}{d}$	$\dfrac{\mu d}{2w}$	$\dfrac{d}{2w}\sqrt{\left(\dfrac{\mu}{\varepsilon}\right)}$

2.2 REFLECTION OF WAVES BY A CONDUCTING PLANE

The simplest case of reflection of waves by a conducting plane occurs when the direction of propagation is normal to the plane. Equation (1.45) shows

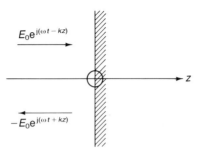

Fig. 2.2 Reflection of electromagnetic waves from a conducting surface at normal incidence.

that the wave impedance in a good conductor tends to zero as ω tends to infinity. The boundary conditions at the surface of the conductor are therefore given by (1.86). If the conducting surface lies in the (x, y) plane as shown in Fig. 2.2 the electric field in the region to the left of the plane is given by

$$E_0 \exp j(\omega t - kz) - E_0 \exp j(\omega t + kz) = -2jE_0 \sin kz \exp(j\omega t). \quad (2.11)$$

The field is therefore a standing wave as shown in Fig. 2.3. The amplitude of the electric field at each point in space is fixed and the actual field values vary sinusoidally with time between the limits indicated by the solid and broken curves shown in the figure.

The corresponding magnetic field is

$$H_0 \exp j(\omega t - kz) + H_0 \exp j(\omega t + kz) = -\frac{2E_0}{Z_0} \cos kx \exp j\omega t. \quad (2.12)$$

This field is also a standing wave but with a maximum rather than a zero of field at the surface of the plane. The factor j which appears in (2.11) but not in (2.12) shows that the electric and magnetic fields are in phase quadrature so that there is no flow of energy. The energy is stored alter-

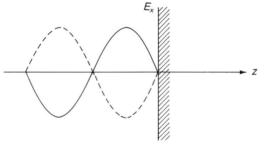

Fig. 2.3 Standing wave produced by the reflection of electromagnetic waves by a conducting surface at normal incidence.

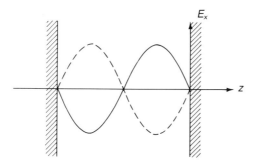

Fig. 2.4 Standing wave produced by reflection of an electromagnetic wave between two parallel conducting planes.

nately in the electric and magnetic fields. This situation is analogous to the transfer of energy between kinetic and potential energy in a pendulum, or between electric and magnetic stored energy in a resonant circuit. A simple electromagnetic resonator can be constructed by putting a second plane parallel to the first. If the second plane is located at one of the zeroes of the electric field, shown in Fig. 2.4, then the boundary conditions will be satisfied upon it. Taking the separation between the planes as d the condition for resonance is

$$\sin kd = 0$$

$$k = \frac{\pi}{d}, \frac{2\pi}{d}, \frac{3\pi}{d}, \text{etc.} \tag{2.13}$$

We shall return to this subject in chapter 7.

2.3 TRANSVERSE ELECTRIC WAVES

When the waves are incident obliquely upon the boundary the situation is rather more complicated. The cases when the electric and magnetic field vectors are parallel to the plane must be treated separately. Any general case can then be regarded as a superposition of these two particular cases.

Consider first the case with the electric field vector parallel to the conducting plane. Equation (1.95) shows that the incident and reflected waves must be in antiphase at the boundary because the tangential component of the electric field must always be zero there. It is helpful to discuss the wave pattern produced by this reflection by thinking in terms of wavefronts. Figure 2.5 shows the electric fields of the incident and reflected waves diagrammatically at an instant of time. The solid lines represent places where the electric field is a maximum and directed out of the paper and the broken lines places where it is maximum but directed into the paper. The directions of motion of these wavefronts are shown by the arrows. At the

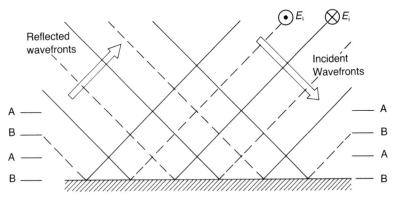

Fig. 2.5 Reflection of a plane electromagnetic wave by a conducting plane for oblique incidence with the electric field vectors parallel to the plane.

conducting surface the incident and reflected waves are in antiphase as required by the boundary conditions.

On planes such as A–A the two waves are in phase with each other producing a maximum of the electric field. This maximum is just twice the amplitude of the incident wave. Similarly on planes such as B–B the waves are in antiphase so that the electric field is zero. These planes are parallel to the conducting surface, are evidently equally spaced, and their positions are independent of the instantaneous positions of the wavefronts. Thus in the direction normal to the plane there is a standing wave exactly as in the case of normal incidence.

The separation between the planes can be found by considering Fig. 2.6. PP' and QQ' are successive wavefronts in a wave with angle of incidence θ. The perpendicular distance between them is equal to the wavelength of the incident wave λ_0. The wavelength along the plane is therefore

$$PQ = \lambda_0/\sin \theta. \qquad (2.14)$$

If Q–Q'' is a reflected wavefront then PP' and QQ'' intersect on the line

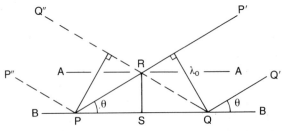

Fig. 2.6 Geometry of the reflection of an electromagnetic wave by a conducting plane for oblique incidence.

AA at R and RS is perpendicular to PQ. Then from the triangle PRS we have

$$RS = PS \tan \theta$$

$$= \lambda_0/(2 \cos \theta) \tag{2.15}$$

which is the separation between the planes A–A and B–B.

Looking again at Fig. 2.5 we see that the pattern of wavefronts is moving parallel to the plane. The wavelength in this direction is $\lambda_0/\sin \theta$ so that the phase velocity is given by

$$v_p = f\lambda_0/\sin \theta = v_{p0}/\sin \theta \tag{2.16}$$

where v_{p0} is the phase velocity of the incident wave. If the space in which the waves are propagating is empty then (2.16) shows that the phase velocity of the waves exceeds the velocity of light. This apparently contradicts the assumption in the theory of relativity that nothing can move faster than the speed of light. There is, in fact, no contradiction as will be shown later on.

The behaviour of the magnetic field can be deduced by considering Fig. 2.8 which shows wavefronts in terms of the magnetic field vectors at a moment in time. The arrows show the directions of the field. The component of H parallel to the surface is not reversed by the reflection so that it is consistent with equation (1.96). The planes A–A and B–B are identical to those in Fig. 2.5. The pattern of the magnetic field is the superposition of the fields of the incident and reflected waves. Using the Cartesian coordinate axes shown in Fig. 2.8 we can resolve the magnetic field vectors parallel and perpendicular to the plane to give

The behaviour of the magnetic field can be deduced by considering Fig. 2.8 which shows wavefronts in terms of the magnetic field vectors at a moment in time. The arrows show the directions of the field. The component of H parallel to the surface is not reversed by the reflection so that it is consistent with equation (1.96). The planes A–A and B–B are identical to those in Fig. 2.5. The pattern of the magnetic field is the

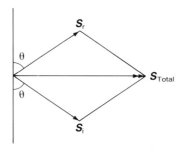

Fig. 2.7 Vector addition of the Poynting vectors of the incident and reflected waves.

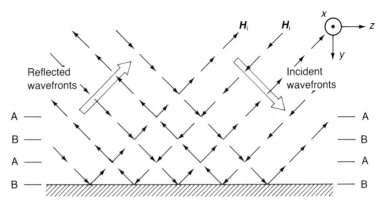

Fig. 2.8 Magnetic field vectors for the reflection of electromagnetic waves by a plane as illustrated in Fig. 2.5.

superposition of the fields of the incident and reflected waves. Using the Cartesian coordinate axes shown in Fig. 2.8 we can resolve the magnetic field vectors parallel and perpendicular to the plane to give

$$H_y = |H_i| \sin \theta \quad \text{and} \quad H_z = -|H_i| \cos \theta \qquad (2.18)$$

for the incident wave, and

$$H_y = -|H_i| \sin \theta \quad \text{and} \quad H_z = -|H_i| \cos \theta \qquad (2.19)$$

for the reflected wave. Then, on planes (A–A) on which the magnetic fields are in antiphase, the z components cancel and

$$H_y = 2|H_i| \sin \theta. \qquad (2.20)$$

Similarly on planes where they are in phase (B–B) the y components cancel so that

$$H_z = 2|H_i| \cos \theta. \qquad (2.21)$$

The local directions of the magnetic field are therefore as shown in Fig. 2.9. The magnetic flux lines are seen to form closed loops as required by equation (1.2).

We are now in a position to write down the equations which define the electric and magnetic field vectors at every point in space. These are

$$E_x = 2|E_i| \sin k_y y \cos (\omega t - k_z z) \qquad (2.22)$$

$$H_y = \frac{2|E_i|}{Z_0} \sin \theta \sin k_y y \cos (\omega t - k_z z) \qquad (2.23)$$

$$H_z = \frac{2|E_i|}{Z_0} \cos \theta \cos k_y y \sin (\omega t - k_z z), \qquad (2.24)$$

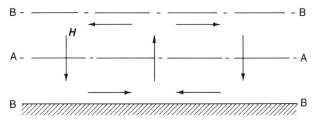

Fig. 2.9 Magnetic field vectors at various positions close to the conducting plane of Fig. 2.5.

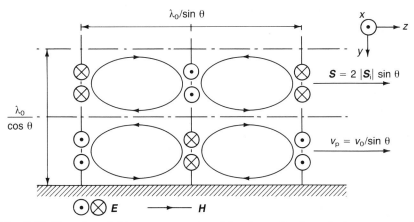

Fig. 2.10 Complete field pattern produced by the reflection of electromagnetic waves by a conducting plane when the electric field vectors are parallel to the plane.

where
$$k_y = \frac{2\pi}{\lambda_0} \cos \theta = k_0 \cos \theta \tag{2.25}$$

and
$$k_z = \frac{2\pi}{\lambda_0} \sin \theta = k_0 \sin \theta. \tag{2.26}$$

These equations show that we can think of a wave travelling at an angle to the coordinate axes as propagating as

$$\exp j(\omega t - k_y y - k_z z), \tag{2.27}$$

where k_y and k_z are the components of the vector propagation constant k_0 in the y and z directions. k_0 is a vector in the direction of propagation whose magnitude is $2\pi/\lambda_0$.

Figure 2.10 summarizes these results in the form of a diagram. The field pattern shown is repeated periodically in both the y and z directions. It is worth spending some time studying Fig. 2.10 in relation to equations (2.22) to (2.24). Notice that H_z is in phase quadrature with E_x so the Poynting vector derived from them has a time average value of zero. This is another way of showing that the power flow is in the z direction.

The wave pattern shown in Fig. 2.10 has magnetic field components both parallel to and perpendicular to the direction of propagation of the wave. The electric field, however, only has a component perpendicular to the z axis. This kind of wave is known as a Transverse Electric (TE) wave. It is useful to distinguish it as a separate kind of wave although it is, in reality, just the superposition of two TEM waves travelling at an angle to each other. Because the magnetic field has a component in the z direction this kind of wave is sometimes referred to as an **H** wave.

2.4 TRANSVERSE MAGNETIC WAVES

When the incident waves are polarized so that the magnetic field vector is parallel to the reflecting surface a somewhat different pattern exists. In this case the boundary conditions to be applied are (1.92) and (1.93) so that the incident and relected waves have their magnetic field vectors in phase at the boundary and the tangential components of the electric field are in antiphase. The analysis of this situation follows exactly the same path as that in the previous section. It will not be followed in detail here but left as an exercise for the student.

By analogy with Fig. 2.5 we can draw a diagram showing the incident and reflected wavefronts but this time in terms of the magnetic field. The result is shown in Fig. 2.11. Note that the waves are reflected without a change of phase. As before there is a standing wave in the direction normal to the plane with planes A–A and B–B on which the fields cancel or add

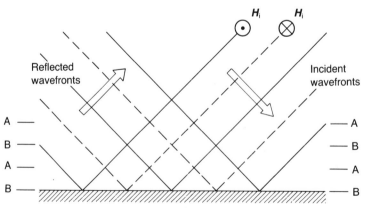

Fig. 2.11 Reflection of a plane electromagnetic wave by a conducting plane for oblique incidence with the magnetic field vectors parallel to the plane.

but this time A–A is a null plane. The separations of the planes can be shown to be exactly as before.

Figure 2.12 shows the electric field vectors by analogy with Fig. 2.8. Notice how the directions of the vectors at the reflecting surface ensure that the tangential component of the electric field is zero. Figure 2.13 shows the field vectors at certain points obtained by superimposing the incident and relected waves.

Finally Fig. 2.14 shows the complete field pattern for this case. Notice how the electric field lines form closed loops except immediately adjacent to the conducting plane where they terminate on surface charges. The equations describing the fields are found by inspection to be

$$H_x = \frac{2|E_i|}{Z_0} \cos (k_y y) \cos (\omega t - k_z z) \tag{2.28}$$

$$E_y = -2|E_i| \sin \theta \cos (k_y y) \cos (\omega t - k_z z) \tag{2.29}$$

$$E_z = 2|E_i| \cos \theta \sin (k_y y) \sin (\omega t - k_z z), \tag{2.30}$$

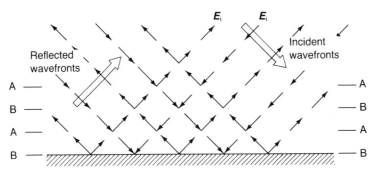

Fig. 2.12 Electric field vectors for the reflection of electromagnetic waves by a plane as illustrated in Fig. 2.11.

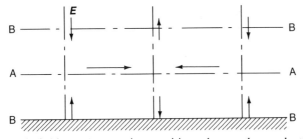

Fig. 2.13 Electric field vectors at various positions close to the conducting plane of Fig. 2.11.

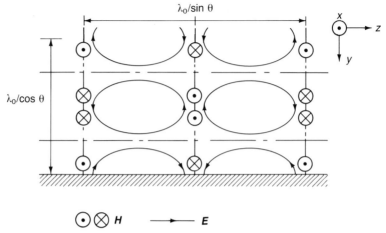

Fig. 2.14 Complete field pattern produced by the reflection of electromagnetic waves by a conducting plane when the magnetic field vectors are parallel to the plane.

where k_y and k_z are defined by (2.25) and (2.26) as before. In this field pattern the magnetic field only has a component perpendicular to the direction of propagation. The wave is accordingly referred to as a transverse magnetic (TM) wave. The alternative terminology, an E wave, is sometimes used.

2.5 PROPAGATION IN A RECTANGULAR WAVEGUIDE

An important practical waveguiding system is the rectangular waveguide shown in Fig. 2.15. This waveguide is simply a brass or copper pipe having a rectangular cross section. Rectangular waveguides are normally made in standard sizes with width a approximately twice the height b. Table 2.2 lists the sizes in common use with their designations, inside dimensions and recommended frequency bands. The reasoning behind the choice of dimensions is discussed in Section 2.7.

Because a rectangular waveguide has only one conductor it cannot support a TEM wave. This is clear from the discussion in Section 2.1 where it was shown that the fields of such a wave satisfy Laplace's equation in planes perpendicular to the direction of propagation. No electrostatic field can exist within a closed conducting boundary which encloses no free charge and therefore TEM waves cannot exist in a rectangular waveguide.

It is, however, possible for both TE and TM waves to propagate down the guide. In this section we shall consider the TE mode which has its electric field vector parallel to the narrow wall of the guide. Other possible modes of propagation will be considered in Section 2.7. To consider how

Table 2.2 Rectangular waveguides in common use

| Designation | | a | b | Frequency | Power |
UK	USA	(mm)	(mm)	(GHz)	(MW)
WG6	WR650	165.1	82.6	1.14 to 1.73	13.5
WG8	WR430	109.2	54.6	1.72 to 2.61	5.9
WG10	WR284	72.14	34.04	2.60 to 3.95	2.4
WG12	WR187	47.6	22.1	3.94 to 5.99	1.0
WG14	WR137	34.85	15.80	5.38 to 8.17	0.54
WG16	WR90	22.86	10.16	8.20 to 12.50	0.23
WG18	WR62	15.80	7.90	11.9 to 18.00	0.12
WG20	WR42	10.67	4.32	17.6 to 26.7	0.048
WG22	WR28	7.11	3.56	26.4 to 40.1	0.025

this TE mode propagates we refer to Fig. 2.10. In this figure the waves are propagating from left to right parallel to the conducting plane and the electric field is normal to the plane of the paper. If another conducting plane is placed parallel to the first so that it coincides with the first null plane of the field pattern then the field between the two planes is unaffected and remains exactly as shown in Fig. 2.10. If, in addition, another pair of conducting planes is placed parallel to the plane of the paper the wave can still propagate because the electric field is perpendicular to these planes and the boundary conditions can be satisfied. The set of four planes so defined form a rectangular conducting pipe exactly as illustrated in Fig. 2.15.

The propagating wave can be thought of as a combination of TEM waves bouncing from wall to wall down the waveguide.

Comparison of Fig. 2.10 and 2.15 shows that

$$\cos\theta = \lambda_0/2a. \tag{2.31}$$

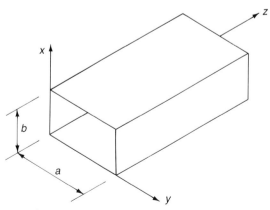

Fig. 2.15 A rectangular waveguide.

It is convenient to define the guide wavelength of the TE wave by

$$\lambda_g = \lambda_0/\sin\theta \qquad (2.32)$$

(see Fig. 2.10) so that

$$\sin\theta = \lambda_0/\lambda_g. \qquad (2.33)$$

The angle θ is not apparent to the user of the waveguide. It can be eliminated by squaring and adding (2.31) and (2.33) to give

$$\frac{1}{\lambda_0^2} = \frac{1}{(2a)^2} + \frac{1}{\lambda_g^2} \qquad (2.34)$$

or

$$\frac{1}{\lambda_g^2} = \frac{1}{\lambda_0^2} - \frac{1}{(2a)^2}. \qquad (2.35)$$

Multiplying this equation by $4\pi^2$, recalling that the propagation constant is given by $k = 2\pi/\lambda$, and that in free space $k = \omega/c$, where c is the velocity of light, we obtain

$$k_g = \frac{1}{c}(\omega^2 - \omega_c^2)^{\frac{1}{2}} \qquad (2.36)$$

where $\omega_c = (\pi c/a)$. To understand the significance of this equation it is helpful to consider the graph of ω against k derived from it shown in Fig. 2.16. When $\omega = \omega_c$, $k_g = 0$ so that the wave does not propagate. The guide is then said to be at cut-off and ω_c is referred to as the cut-off frequency. The physical significance of this result can be shown by considering equation (2.33). At cut-off λ_g tends to infinity so that $\theta = 0$ and the wave bounces backwards and forwards across the guide without progressing down it.

At frequencies above cut-off the relationship between the guide wavelength and the frequency is given by the curve in Fig. 2.16. At any frequency

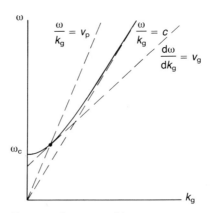

Fig. 2.16 Dispersion diagram of a waveguide.

the phase velocity of the wave is given by the ratio of the frequency to the propagation constant, that is by the slope of the line from a point on the curve to the origin of the graph. Mathematically

$$v_p = \frac{\omega}{k_g} = c\left[1 + \left(\frac{\pi}{k_g a}\right)^2\right]^{\frac{1}{2}}. \tag{2.37}$$

Thus the phase velocity of a wave in an empty waveguide is always greater than the velocity of light and tends to c as k_g tends to infinity. Unlike the TEM transmission lines discussed earlier the phase velocity of a wave varies with frequency. This means that, if an electromagnetic pulse is injected into the guide, the different Fourier components from which it is made up travel with different velocities and the shape of the pulse which emerges from the end of the guide differs from that injected. The pulse is said to be *dispersed* by the guide and the guide itself is described as dispersive. This property is not limited to rectangular waveguides, it occurs in any wave propagating medium in which the phase velocity is a function of frequency. Figure 2.16 is called the dispersion diagram of the waveguide.

Equation (2.37) apparently violates the axiom of the theory of relativity that nothing can travel faster than the velocity of light. There is in fact no contradiction. The axiom is more accurately stated as: 'Information cannot travel faster than the velocity of light.' A continuous sine wave carries no information, to become a carrier of information it must be modulated in some way. The effects of modulation are most easily illustrated by considering the superposition of two signals having the same amplitude and slightly different frequencies. Technically this is double-sideband suppressed-carrier modulation (J.J. O'Reilly, 1984). If the two waves are

$$E_1 = E_0 \exp j[(\omega + \delta\omega)t - (k + \delta k)z]$$

and
$$E_2 = E_0 \exp j[(\omega - \delta\omega)t - (k - \delta k)z]. \tag{2.38}$$

These expressions can be rearranged to give

$$E_1 = E_0 \exp j(\omega t - kz) \exp j(\delta\omega t - \delta kz)$$

and
$$E_2 = E_0 \exp j(\omega t - kz) \exp [-j(\delta\omega t - \delta kz)]. \tag{2.39}$$

When they are added together the result is

$$E = 2E_0 \exp j(\omega t - kz) \cos (\delta\omega t - \delta kz). \tag{2.40}$$

This expression can be interpreted as a carrier wave which propagates as $\exp j(\omega t - kz)$ whose amplitude is $2E_0 \cos (\delta\omega t - \delta kz)$. The envelope of the carrier wave carries information at a velocity given by

$$v_g = \lim_{\delta\omega \to 0} \left(\frac{\delta\omega}{\delta k}\right) = \frac{d\omega}{dk}. \tag{2.41}$$

This velocity is known as the *group velocity* of the wave. On the dispersion diagram (Fig. 2.16) it is represented by the slope of the tangent to the dispersion curve at a point. It is evidently always less that the velocity of light and it tends to zero at the cut-off frequency.

By differentiating (2.36) it can be shown that the group velocity in a rectangular waveguide is

$$v_g = c(\lambda_0/\lambda_g) = c \sin \theta \tag{2.42}$$

which is just the z component of the phase velocity of the wave bouncing down the guide.

2.6 POWER FLOW IN A RECTANGULAR WAVEGUIDE

The fields of the TE mode discussed in the previous section are readily derived from equations (2.22) to (2.24) since

$$k_y = \pi/a \tag{2.43}$$

from (2.25) and (2.31), and

$$k_z = k_g. \tag{2.44}$$

So, making use of (2.31) and (2.33)

$$E_x = E_0 \sin \left(\frac{\pi y}{a}\right) \cos (\omega t - k_g z) \tag{2.45}$$

$$H_y = \frac{E_0}{Z_0} \frac{\lambda_0}{\lambda_g} \sin \left(\frac{\pi y}{a}\right) \cos (\omega t - k_g z) \tag{2.46}$$

$$H_z = \frac{E_0}{Z_0} \frac{\lambda_0}{\lambda_c} \cos \left(\frac{\pi y}{a}\right) \sin (\omega t - k_g z). \tag{2.47}$$

From these equations it can be seen that E_x and H_y are in phase with one another whilst H_z has a phase difference from them of 90°. The time average Poynting vector is therefore

$$S_z = \tfrac{1}{2}|E_x| |H_y| \tag{2.48}$$

and the average power flow is obtained by integrating S_z across the cross section of the guide

$$W = \frac{1}{2} \int_0^b dx \int_0^a |E_x| |H_y|\, dy$$

$$= \frac{ab}{4} \frac{\lambda_0}{\lambda_g} \frac{|E_0|^2}{Z_0}. \tag{2.49}$$

This equation shows that the power flowing in a guide depends upon the electric field strength at the centre of the guide and the cross-sectional area. In a guide filled with air at atmospheric pressure the electric field is limited by dielectric breakdown. This sets a limit on the power density in the guide. The maximum power handling capabilities of waveguides decrease with the square of the cut-off frequency. Table 2.2 shows the maximum recommended power-handling capabilities of standard waveguides. Higher powers can be dealt with by pressurizing the guide either with air or with a gas such as freon or sulphur hexafluoride. The power which can be trans- mitted can also be increased by evacuating the guide because that also raises the breakdown field.

By calculating the total stored energy in one wavelength of the wave and comparing it with (2.48) it is possible to show that the velocity of propagation of energy is equal to the group velocity (see Exercise 2.3).

If the signal frequency is below the cut-off frequency then k is imaginary as can be seen from (2.36). The wave no longer propagates down the guide and the fields decay as exp $-k_g z$. Equations (2.45) and (2.46) show that E_x and H_y are in phase quadrature (because λ_g is imaginary) so that the Poynting vector is purely imaginary. There is then no flow of energy but only a reactive storage of it. A cut-off wave of this kind is known as an evanescent wave.

It is often useful to apply transmission line theory to waveguide prob- lems. To do this we require the characteristic impedance of the guide as well as the guide wavelength. The characteristic impedance is defined by

$$Z_g = |V_0|^2 / 2W, \tag{2.50}$$

where $V_0 = bE_0$ is the magnitude of the potential difference between the centres of the broad walls of the guide. Substituting for the power flow from (2.49) yields

$$Z_g = \frac{2b}{a} \frac{\lambda_g}{\lambda_0} Z_0. \tag{2.51}$$

Note that the characteristic impedance of a waveguide varies with frequency.

The theory of transmission lines is commonly based upon an equivalent circuit for the line. Our discussion of rectangular waveguides has, so far, adopted a field approach. It is, however, possible to represent a waveguide by an equivalent circuit. The derivation of that circuit provides a useful example of the way in which such circuits can be derived for microwave components.

Figure 2.17(a) shows a sectioned view of a waveguide including the electric and magnetic field patterns. To satisfy the boundary conditions there must be charge concentrations and currents flowing in the walls as shown in Fig. 2.17(b). To derive the equivalent circuit of a short length dz of the guide we include shunt capacitance to represent the displacement current paths and shunt and series inductance to represent the conduction current paths. The result is the circuit shown in Fig. 2.18. In order to derive

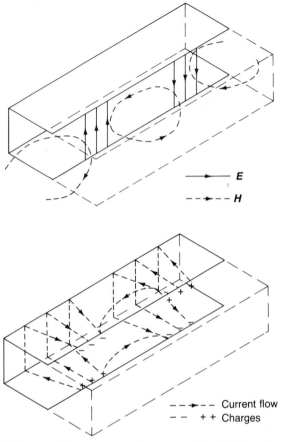

$$\longrightarrow \quad E$$
$$- - \cdot \longrightarrow - - \cdot \quad H$$

$$- - \cdot \longrightarrow \cdot - - \quad \text{Current flow}$$
$$- - \quad + + \quad \text{Charges}$$

Fig. 2.17 The TE_{01} mode in a rectangular waveguide: (a) Electric and magnetic field patterns, and (b) wall charges and currents.

expressions for the component values in terms of the dimensions of the guide we require three conditions, namely:

1. The equivalent circuit must have the same cut-off frequency as the guide;
2. The phase velocity calculated from the equivalent circuit must tend to the velocity of light in the limit of high frequencies; and
3. The equivalent circuit must have the same characteristic impedance as the guide.

Assuming that the phase change produced by the section is $k\,dz$ the analysis of the circuit shows that

$$k = \left[\frac{L_2}{L_1} (\omega^2 L_1 C - 1) \right]^{\frac{1}{2}}. \tag{2.52}$$

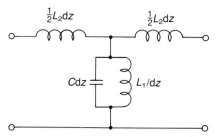

Fig. 2.18 Equivalent circuit for the TE_{01} mode in a rectangular waveguide.

This equation fits exactly the curve shown in Fig. 2.16 provided that

$$L_1 C = 1/\omega_c^2 \quad \text{and} \quad \frac{L_2}{L_1} = \frac{\omega_c^2}{c^2} \tag{2.53}$$

which satisfy conditions 1 and 2. The iterative impedance of the network tends to

$$Z_g = \frac{\omega L_1}{\omega^2 L_1 C - 1} k \tag{2.54}$$

as dz tends to zero. Equating the expressions for Z_g given by (2.51) and (2.54) to satisfy condition 3 gives

$$L_1 = \frac{2ab}{\pi^2} \mu_0 \tag{2.55}$$

and, from (2.53) we obtain

$$C = \frac{a}{2b} \varepsilon_0 \tag{2.56}$$

and $$L_2 = \frac{2b}{a} \mu_0. \tag{2.57}$$

It is important to note that the dimensions of these expressions are henry metres, farads per metre and henries per metre, respectively. It is also important to note that (2.56) is not equal to the capacitance per unit length between the broad walls of the guide regarded as a parallel plate capacitor. This underlines the point that the fields in a waveguide cannot be obtained by solving Laplace's equation.

2.7 HIGHER-ORDER MODES IN A RECTANGULAR WAVEGUIDE

In Section 2.5 we saw that a rectangular waveguide could be constructed by inserting a conducting sheet along the null plane closest to the conducting

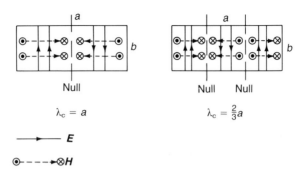

Fig. 2.19 Higher-order modes in rectangular waveguide: (a) TE_{02} and, (b) TE_{03}.

plane in the field pattern shown in Fig. 2.10. We could just as easily have chosen any of the other null planes for this purpose. This would have led to other possible solutions for the propagation of waves in the guide. Figure 2.19 shows the next two such modes with their cut-off wavelengths ($\lambda_c = 2\pi c/\omega_c$). Each has a longitudinal null plane and a transverse electric field described by

$$E_x = E_0 \sin\left(\frac{n\pi y}{a}\right) \cos(\omega t - k_g z), \tag{2.58}$$

where $n = 1, 2, 3$, etc. We could, instead, have chosen to study modes with their electric fields in the y direction so that

$$E_y = E_0 \sin\left(\frac{m\pi x}{b}\right) \cos(\omega t - k_g z), \tag{2.59}$$

where $m = 1, 2, 3$, etc. With such a proliferation of modes it is desirable to have a way of referring to them. The usual notation is to call them TE_{mn} modes where m and n have integer values. On this basis the mode discussed in Sections 2.5 and 2.6 is the TE_{01} mode and those shown in Fig. 2.19 are the TE_{02} and TE_{03} modes. It is natural to ask whether modes can exist for which m and n are both non-zero. A complete solution of the electromagnetic wave equation for transverse electric waveguide modes (Ramo et al., 1965) shows that this is indeed the case. Figure 2.20 shows, for example, the field pattern for the TE_{11} mode. It is useful to be able to sketch field patterns for the different modes because it is then possible to write down the equations which describe them by inspection. Thus from Fig. 2.20 we get

$$E_x = E_1 \cos\left(\frac{\pi x}{b}\right) \sin\left(\frac{\pi y}{a}\right) \cos(\omega t - k_g z) \tag{2.60}$$

and $\qquad E_y = E_2 \sin\left(\frac{\pi x}{b}\right) \cos\left(\frac{\pi y}{a}\right) \cos(\omega t - k_g z).$ $\qquad$ (2.61)

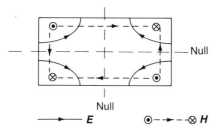

Fig. 2.20 The TE$_{11}$ mode in rectangular waveguide.

This field pattern must satisfy the wave equation (2.4) so that

$$\left(\frac{\pi}{b}\right)^2 + \left(\frac{\pi}{a}\right)^2 + k_g^2 = k_0^2. \tag{2.62}$$

In general for a TE$_{mn}$ mode

$$\left(\frac{m\pi}{b}\right)^2 + \left(\frac{n\pi}{a}\right)^2 + k_g^2 = k_0^2 \tag{2.63}$$

so that all the modes obey the equation

$$k_g^2 = k_0^2 - k_c^2, \tag{2.64}$$

where $$k_c^2 = \left(\frac{2\pi}{\lambda_c}\right)^2 = \left(\frac{m\pi}{b}\right)^2 + \left(\frac{n\pi}{a}\right)^2. \tag{2.65}$$

Equation (2.64) is a generalization of (2.35). It can be seen from (2.65) that λ_c decreases and therefore the cut-off frequency increases as m and n increase. The mode which has the lowest cut-off frequency of all is the TE$_{01}$ mode discussed in Sections 2.5 and 2.6. Table 2.3 shows the cut-off frequencies for some of the lower modes in standard WG16 waveguide.

Table 2.3 Cut-off frequencies of TE modes in waveguide WG16

| Mode | | Cut-off frequency |
m	n	(GHz)
0	1	6.55
0	2	13.10
1	0	14.71
1	1	16.10
0	3	19.65
1	2	19.69
1	3	24.54
2	1	30.13
2	2	32.20

The relationship between the amplitudes of the x and y components of the electric field can be obtained by making use of Maxwell's equation (1.5) which is

$$\frac{\partial E_x}{\partial x} + \frac{\partial E_y}{\partial y} = 0 \qquad (2.66)$$

since there is no free charge within the waveguide and $E_z = 0$. Substitution from (2.60) and (2.61) yields

$$\left[\frac{\pi E_1}{b} + \frac{\pi E_2}{a}\right]\left[\sin\left(\frac{\pi x}{b}\right)\sin\left(\frac{\pi y}{a}\right)\cos\left(\omega t - k_g z\right)\right] = 0 \qquad (2.67)$$

so that

$$E_2 = -\frac{a}{b}E_1. \qquad (2.68)$$

A little thought shows that this result is equivalent to saying that the electric field lines must all start and finish on the walls of the guide. This means that the amplitudes of the fields at the centres of the walls must be in inverse ratio to the widths of the walls.

The magnetic field pattern of a mode can be obtained in a similar way by substituting the electric field into Maxwell's equation (1.8). For the TE_{11} mode the result is

$$-j\omega B = \begin{vmatrix} \hat{x} & \hat{y} & \hat{z} \\ \partial/\partial x & \partial/\partial y & \partial/\partial z \\ E_x & E_y & 0 \end{vmatrix}$$

$$= \left[-\frac{\partial E_y}{\partial z}\right]\hat{x} + \left[\frac{\partial E_x}{\partial z}\right]\hat{y} + \left[\frac{\partial E_x}{\partial y} - \frac{\partial E_y}{\partial z}\right]\hat{z}. \qquad (2.69)$$

Substituting the expressions for the electric fields gives

$$H_x = \frac{jk_g}{\omega\mu_0}\frac{a}{b}E_1\sin\left(\frac{\pi x}{b}\right)\cos\left(\frac{\pi y}{a}\right)\sin\left(\omega t - k_g z\right) \qquad (2.70)$$

$$H_y = \frac{jk_g}{\omega\mu_0}E_1\cos\left(\frac{\pi x}{b}\right)\sin\left(\frac{\pi y}{a}\right)\sin\left(\omega t - k_g z\right) \qquad (2.71)$$

and

$$H_z = \frac{j}{\omega\mu_0}\left[\frac{\pi E_1}{a} - \frac{\pi a E_1}{b^2}\right]\cos\left(\frac{\pi x}{b}\right)\cos\left(\frac{\pi y}{a}\right)\cos\left(\omega t - k_g z\right). \qquad (2.72)$$

These expressions can be derived by solving the wave equation (2.3) subject to the appropriate boundary conditions but the approach given here has the advantage of being tied more closely to an intuitive understanding of electromagnetic phenomena.

So far we have only considered transverse electric modes. Section 2.4 demonstrated the existence of transverse magnetic modes when plane waves are incident obliquely on a conducting plane. Can such waves pro-

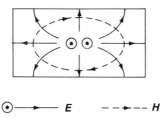

$\odot$———— E $--\!\!\rightarrow\!\!--$ H

Fig. 2.21 The TM_{11} mode in rectangular waveguide.

pagate in a waveguide? The answer is that they can. Like the TE waves they are actually the result of the superposition of plane TEM waves. Unfortunately there is no simple construction like that used for TE waves by which the TM mode patterns can be deduced. The problem is that although a conducting sheet can be introduced parallel to the conducting plane in Fig. 2.14 it is not possible simply to add a further pair of planes at right angles to the first. This is because the tangential component of the electric field cannot be zero on them.

The full solution to the wave equation for the TM modes can be found in standard texts (Ramo *et al.*, 1965). Here we shall proceed by sketching the field patterns and observing that the fields can be deduced from them as before. Figure 2.21 shows the field pattern for the TM_{11} mode. This is, in fact, the lowest frequency TM mode because modes with m or $n = 0$ cannot satisfy the boundary conditions. It is easy to show that the TM modes must also satisfy equations (2.64) and (2.65) so that their behaviour is just like that of the TE modes.

We are now in a position to explain the operating bands quoted for standard waveguides in Table 2.2. Operation in the TE_{01} mode ensures that that is the only mode which can propagate; all the others are cut off. As the frequency approaches the cut-off frequency the guide becomes increasingly dispersive. From equation (2.42) we find that the group velocity is

$$v_g = [1 - (\omega_c/\omega)^2]^{\frac{1}{2}}c. \tag{2.73}$$

At the bottom of the frequency band quoted for a waveguide the group velocity is 60% of the velocity of light.

The upper frequency limit is set by the need to ensure that all the higher order modes are well beyond cut-off. Table 2.3 shows that cut-off frequency of the TE_{02} mode in waveguide 16 is a factor of 1.05 above the highest working frequency of the guide. From equation (2.64) we find that the propagation constant is $82j\,m^{-1}$. At the same frequency the propagation constant of the TE_{01} mode is $223\,m^{-1}$. The attenuation of the evanescent TE_{02} mode signal in one wavelength of the TE_{01} mode is $46\,dB$.

The useful bandwidth of a rectangular waveguide is therefore governed by the separation between the cut-off frequencies of the lowest and next to lowest modes of propagation. The aspect ratio of $2:1$ which is commonly used makes the TE_{10} cut-off frequency a little higher than that of the TE_{02} mode. It is therefore close to the ratio of dimensions which gives the maximum possible bandwidth.

2.8 OTHER WAVEGUIDES

Although all the discussion so far in this chapter has been about rectangular waveguides it is obvious that any conducting pipe can act as a waveguide with properties similar to those of the rectangular guides. Two which are commonly encountered are the circular and ridge waveguides.

The circular waveguide has the special property that no plane containing the axis is distinguishable from any other. As a result different polarizations of a TE mode have the same cut-off and guide wavelengths. This property is employed in the rotating joints used in the waveguides feeding radar antennas. It is also used to make accurate attenuators and phase shifters as we shall see in Chapter 4.

The analysis of modes in circular waveguide requires the use of cylindrical polar coordinates (see Appendix B). The suffixes which describe the modes of propagation refer to the number of nodes in the tangential (θ) and radial (r) directions, respectively. Figure 2.22 shows a few of the modes which exist in a cylindrical waveguide. Details of the fields and the cut-off wavelengths for the different modes can be found in Ramo *et al.* (1965).

In the previous section we saw that the useful bandwidth of a rectangular waveguide is limited to around $1.5:1$. For some purposes this is inconvenient. In order to increase the useful bandwidth we need to increase the

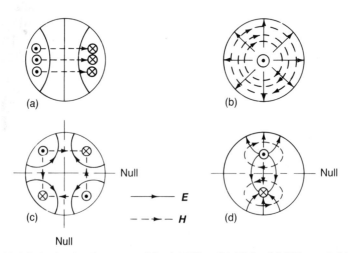

Fig. 2.22 Modes in circular waveguide: (a) TE_{11}, (b) TM_{01}, (c) TE_{21} and, (d) TM_{11}.

Fig. 2.23 Single- and double-ridge waveguides.

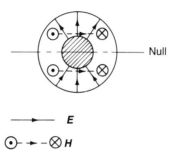

Fig. 2.24 The TE_{11} waveguide mode in a coaxial line.

separation between the cut-off frequencies of the TE_{01} and TE_{02} modes. The equivalent circuit shown in Fig. 2.18 suggests a possible way. If a ridge is added to the centre of one, or both, of the broad walls of the guide, as shown in Fig. 2.23, then the effect on the TE_{01} mode will be to increase both the shunt capacitance and the shunt inductance shown in Fig. 2.18. The consequence is that the cut-off frequency of this mode is lowered. The TE_{02} mode, however, will be little affected because the ridges are in regions of weak electric field. The price paid for the increase in bandwidth is a reduction in power handling capability. At low power levels two-wire lines provide a more compact means of broadband signal transmission, but ridge waveguides are useful where a combination of broad-bandwidth and moderately high power is required. The notation for modes in ridge wave-guide is the same as that used for rectangular waveguides.

It is not always appreciated that two-wire lines can also support higher-order modes of propagation. Figure 2.24 shows the TE_{11} mode for a coaxial line. Like any other waveguide mode this has a lower cut-off frequency and a guide wavelength which obeys equation (2.64). The useful bandwidth of a coaxial line is much greater than that of a waveguide because the lower cut-off frequency is zero. The presence of higher-order modes provides a limit on the highest frequency for which a line can be used.

2.9 CONCLUSION

In this chapter we have seen how electromagnetic waves are guided by conducting surfaces. The TEM waves which propagate in free space combine when reflected off a conductor to give waves which can be classified as either TE or TM. Waves of both these types can propagate down metallic tubes at frequencies above some cut-off frequency. The phase velocities of these waves vary with frequency so that pulses formed of groups of waves of different frequencies are dispersed as they travel down the waveguide. The phase velocity is always greater than the velocity of light but it is found that information and power propagate with the group velocity which is less than the velocity of light.

Waveguides can support an infinite set of higher-order modes. The useful bandwidth of a waveguide is limited by the need to avoid excessive dispersion on the one hand and multi-mode propagation on the other. Rectangular waveguides are most commonly used but other cross sections including circular and ridge waveguides are valuable for special purposes. Two-wire transmission lines can also support higher-order modes.

EXERCISES

2.1 Calculate the dimensions of the following $50\,\Omega$ transmission lines:

    1. Coaxial cable with polythene dielectric ($\varepsilon_r = 2.25$) and an inner conductor diameter of 1.5 mm.

    2. An air-spaced parallel-wire line with a conductor diameter of 5 mm.

    3. A triplate line with alumina dielectric ($\varepsilon_r = 8.9$) and centre conductor width 2.5 mm.

2.2 Calculate the cut-off wavelength of the WG10 waveguide and the guide wavelength at 2.5, 3.0, 3.5 and 4.0 GHz. Plot a dispersion diagram for the waveguide and compute the phase and group velocities at 3.0 GHz.

2.3 Find an expression for the stored energy per wavelength in a rectangular waveguide and show, by comparison with (2.49) that energy propagates down the guide at the group velocity.

2.4 Calculate the cut-off frequencies of the TE_{02}, TE_{11} and TM_{11} modes in a WG10 waveguide.

2.5 Calculate the characteristic impedances of waveguides whose heights are 2.0, 4.0, 6.0 and 7.9 mm and whose width is 15.8 mm.

2.6 Calculate the cut-off frequency and characteristic impedance of WG12 waveguide when it is filled with: 1. air at atmospheric pressure and, 2. paraffin wax ($\varepsilon_r = 2.25$).

Waves with dielectric boundaries

3.1 REFLECTION OF WAVES BY A DIELECTRIC BOUNDARY

In Section 1.9 we saw that when waves encounter a boundary there are, in general, both reflected and transmitted waves. This is in accord with the everyday experience that sunlight is partly reflected and partly transmitted by a window. Engineers commonly speak of the transmission loss and return loss which are expressed in decibels as

$$\text{Transmission loss} = -10 \log_{10} \left| \frac{S_t}{S_i} \right| \tag{3.1}$$

and
$$\text{Return loss} = -10 \log_{10} \left| \frac{S_r}{S_i} \right| \tag{3.2}$$

where S_i, S_t and S_r are the Poynting vectors of the incident, transmitted and reflected waves. The Poynting vector in a lossless dielectric material is related to the electric field strength and the wave impedance by

$$|S| = \frac{1}{2}|E|^2/Z_w = \frac{1}{2} \sqrt{\left(\frac{\varepsilon}{\mu}\right)} |E|^2. \tag{3.3}$$

For the case of normal incidence on a dielectric surface the amplitudes of the transmitted and reflected waves can be calculated from (1.84) and (1.85) with the results

$$\frac{E_t}{E_i} = \frac{2\sqrt{\varepsilon_1}}{\sqrt{\varepsilon_1} + \sqrt{\varepsilon_2}}$$

$$\frac{E_r}{E_i} = \frac{\sqrt{\varepsilon_1} - \sqrt{\varepsilon_2}}{\sqrt{\varepsilon_1} + \sqrt{\varepsilon_2}}. \tag{3.4}$$

The permeabilities of all dielectric materials are equal to μ_0 to engineering accuracies so they cancel out of the equations. It is evident that both the wave amplitudes are non-zero except for the trivial case when $\varepsilon_1 = \varepsilon_2$.

3.2 TOTAL INTERNAL REFLECTION

The position is rather different when the angle of incidence is not zero. Equation (1.90) shows that if the permeabilities are assumed to be equal then

$$\sin \theta_2 = \sqrt{\left(\frac{\varepsilon_1}{\varepsilon_2}\right)} \sin \theta_1. \tag{3.5}$$

If ε_1 is greater than ε_2 then there will be a critical value of θ_1 such that $\sin \theta_2 = 1$. For angles of incidence larger than this the wave must be totally reflected. This phenomenon is sometimes observed in underwater photographs where the water surface appears to be a mirror. It has important applications in optical fibres to which we shall return later. At a more mundane level it is employed in the edge-lit signs sometimes seen in cinemas. For the interface between glass and air the critical angle is around 42°.

The phenomenon can also be explained by making use of the concept of wave impedance. From (1.84) and (1.94) the reflected wave amplitude is given by

$$\frac{E_r}{E_i} = \frac{\sqrt{\varepsilon_1} \cos \theta_2 - \sqrt{\varepsilon_2} \cos \theta_1}{\sqrt{\varepsilon_1} \cos \theta_2 + \sqrt{\varepsilon_2} \cos \theta_1} \tag{3.6}$$

when the magnetic field vectors are parallel to the surface. Clearly $|E_r| = |E_i|$ when $\theta_2 = 90°$. Z_{n2} is then zero so this case corresponds to the termination of a transmission line by a short circuit. If the wave is polarized with the electric field vector parallel to the surface we obtain

$$\frac{E_r}{E_i} = \frac{\sqrt{\varepsilon_1} \cos \theta_1 - \sqrt{\varepsilon_2} \cos \theta_2}{\sqrt{\varepsilon_1} \cos \theta_1 + \sqrt{\varepsilon_2} \cos \theta_2} \tag{3.7}$$

using (1.97) and the conclusion is the same except that Z_{n2} is now infinite corresponding to an open-circuit termination.

An alternative, physical, explanation can be obtained by considering Fig. 1.8. If ε_1 is greater than ε_2 the wave is refracted away from the normal. The limit of this process is reached when the refracted wave is moving parallel to the boundary as shown in Fig. 3.1. For angles of incidence greater than the critical angle the separation of the points A and B is less than the wavelength in region 2 and the boundary conditions cannot be satisfied by a propagating wave. Nevertheless the fields in this region must satisfy the wave equation (2.4). Assuming that the fields vary only with y, z and t then, from (2.27)

$$E = E_2 \exp j(\omega t - k_y y - k_z z) \tag{3.8}$$

and, substituting this into (2.4) we get

$$k_y^2 + k_z^2 = k_2^2, \tag{3.9}$$

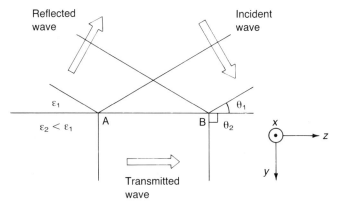

Fig. 3.1 Geometry for total internal reflection of a wave by a boundary between two dielectric materials.

where k_2 is the propagation constant in region 2. If $k_2 < k_z$ then k_y must be imaginary so that the wave decays exponentially away from the boundary. A wave of this kind is called an evanescent wave.

3.3 THE BREWSTER ANGLE

From (3.6) and (3.7) it appears that there could be conditions under which there is no reflection. Taking (3.7) first the condition is

$$\frac{\cos \theta_2}{\cos \theta_1} = \sqrt{\left(\frac{\varepsilon_1}{\varepsilon_2}\right)}. \tag{3.10}$$

If $\varepsilon_2 > \varepsilon_1$ then $\theta_2 > \theta_1$ from (3.10) and $\theta_2 < \theta_1$ from (3.5). These conclusions are contradictory so the condition can never be satisfied. In the other case when the magnetic field is parallel to the boundary we get

$$\frac{\cos \theta_2}{\cos \theta_1} = \sqrt{\left(\frac{\varepsilon_2}{\varepsilon_1}\right)} \tag{3.11}$$

and this time there is no contradiction. From (3.5) and (3.11)

$$\sin^2 \theta_2 + \cos^2 \theta_2 = \frac{\varepsilon_1}{\varepsilon_2} \sin^2 \theta_1 + \frac{\varepsilon_2}{\varepsilon_1} \cos^2 \theta_1 \tag{3.12}$$

or

$$1 = \left[\frac{\varepsilon_1}{\varepsilon_2} - \frac{\varepsilon_2}{\varepsilon_1}\right] \sin^2 \theta_1 + \frac{\varepsilon_2}{\varepsilon_1} \tag{3.13}$$

which yields

$$\sin^2 \theta_1 = \frac{1}{1 + \varepsilon_1/\varepsilon_2}. \tag{3.14}$$

The right-hand side of this equation is less than unity for all possible values of the permittivities and therefore there is always an angle of incidence which produces no reflected wave for this polarization. This angle is known as the Brewster angle. For light in air incident on the surface of water ($\varepsilon_r = 81$) the angle is 83.7°. Even at angles away from the Brewster angle it is to be expected that the reflection coefficients will differ for the different polarizations. If the incident radiation contains equal proportions of both polarizations then the reflected radiation will be partially polarized. This phenomenon is exploited by photographers who use polarizing filters to cut down the intensity of light reflected off water. It is also employed in the output windows of gas lasers to ensure that the light emitted is polarized.

Throughout this section examples of the practical applications of phenomena have been drawn from optics. It is important to remember that they apply equally to the remainder of the electromagnetic spectrum.

3.4 DIELECTRIC WAVEGUIDES

The property of total internal reflection at an interface between two dielectric materials suggests the possibility of using strips or rods to guide waves in a manner similar to the metallic waveguides discussed in the last chapter. There are, however, important differences between the two cases as we shall see. When a wave is reflected from a metal surface the phase change is independent of the angle of incidence. This is no longer the case for total internal reflection from a dielectric boundary.

To investigate this further let us consider waves whose electric field vectors are parallel to the boundary. The amplitude of the reflected wave is given by (3.7). The angle of refraction can be eliminated by noting that

$$\cos \theta_2 = \pm(1 - \sin^2 \theta_2)^{\frac{1}{2}} = \pm\left(1 - \frac{\varepsilon_1}{\varepsilon_2} \sin^2 \theta_1\right)^{\frac{1}{2}}. \qquad (3.15)$$

Now the wave in region 2 propagates in the direction normal to the boundary as $\exp -jk_2 \cos \theta_2 y$. When the angle of incidence exceeds the critical angle the second term on the right hand side of this equation must be greater than unity. $\cos \theta_2$ is then imaginary and the wave decays exponentially with distance from the boundary. Thus it is the negative sign in (3.15) which has physical significance.

Substituting this into (3.7) gives

$$\frac{E_r}{E_i} = \frac{\cos \theta_1 + j(\sin^2 \theta_1 - \varepsilon_2/\varepsilon_1)^{\frac{1}{2}}}{\cos \theta_1 - j(\sin^2 \theta_1 - \varepsilon_2/\varepsilon_1)^{\frac{1}{2}}} \qquad (3.16)$$

so that the phase of the reflected wave is

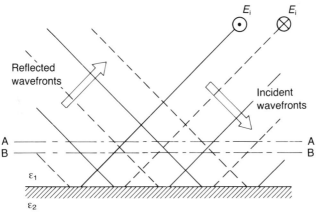

Fig. 3.2 Reflection of plane waves by a boundary between two dielectric materials with the electric field vectors parallel to the boundary.

$$\phi = 2 \tan^{-1} \left[\frac{(\sin^2 \theta_1 - \varepsilon_2/\varepsilon_1)^{\frac{1}{2}}}{\cos \theta_1} \right] \qquad (3.17)$$

relative to the incident wave.

Now consider the interference between plane waves reflected by a dielectric boundary as shown in Fig. 3.2. This figure differs from Fig. 2.5 in that the phase difference between the incident and reflected waves at the boundary is no longer 180°. Within the interference pattern of the waves there are other planes such as A–A on which the phase difference is also ϕ. There are also planes such as B–B on which the phase difference is $-\phi$. If a second boundary coincides with B–B then the boundary conditions are correct for reflection because the roles of the incident and reflected waves have been exchanged with each other.

Figure 3.3 shows the pattern of waves within a dielectric slab of thickness d. The condition for such a pattern to exist is

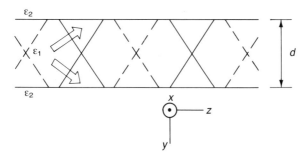

Fig. 3.3 Propagation of waves within a dielectric slab.

$$2k_y d + 2\phi = 2m\pi \tag{3.18}$$

that is, a wave which travels across the slab and back (being reflected twice) must travel a whole number of wavelengths.

Equation (3.18) can be written

$$k_1 d \cos \theta = m\pi - \phi. \tag{3.19}$$

If the phase shift at the boundary is set to π in this equation it is easy to show that it is equivalent to (2.15).

Substituting from (3.18) into an equation equivalent to (3.9) gives the guide equation for a dielectric waveguide

$$k_g^2 = k_1^2 - \left(\frac{m\pi - \phi}{d}\right)^2 \tag{3.20}$$

which can be compared with (2.64). It must be remembered that ϕ is a function of θ_1. At cut-off the angle of incidence is equal to the critical angle so that $\theta_2 = 90°$ in (3.7) and $\phi = 0$ so that

$$k_g = k_1 \sin \theta_1 = k_1 \sqrt{\left(\frac{\varepsilon_2}{\varepsilon_1}\right)} \tag{3.21}$$

and, after a little manipulation the free-space cut-off wavelength is found to be

$$\lambda_c = \frac{2d}{m} \sqrt{\left(\frac{\varepsilon_1}{\varepsilon_0} - \frac{\varepsilon_2}{\varepsilon_0}\right)} \tag{3.22}$$

where $m = 1, 2, 3$, etc.

We see, therefore, that a dielectric slab is capable of supporting a set of modes of propagation very like those in a waveguide. The modes discussed so far are TE modes. Consideration of the other polarity of the waves would lead to TM modes. The difference between the dielectric waveguide modes and those in a metal waveguide is illustrated in Fig. 3.4. In a dielectric guide the waves are not confined to the guide but spread out a little into the surrounding dielectric. The boundaries of the guide therefore lie a little inside those points at which the TEM waves are in antiphase just as shown in Fig. 3.3. At cut-off the wave outside the guide spreads out to infinity so that the cut-off condition is an open-circuit rather than a short-circuit resonance.

We have, so far, avoided discussion of the effects of the lateral boundaries of the guide. The reason is that the situation here is more complicated than it is in metal waveguides. This is clear from Fig. 3.5. A mode which has its electric field in the x direction cannot satisfy the boundary conditions at the dielectric interfaces at $y = 0$ and $y = b$. When the effects of these boundaries are taken into account the patterns of the modes are altered some-

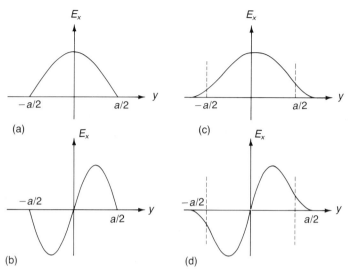

Fig. 3.4 Comparison between TE modes in metal and dielectric waveguides: metal waveguide (a) TE_{01} mode and (c) TE_{02} mode, dielectric waveguide (b) TE_{01} mode and (d) TE_{02} mode.

what but the general conclusion that a set of modes of propagation exists is unaffected. If the strip of dielectric is made much wider than its thickness ($b \ll a$) then the theory given above may be expected to be a reasonable approximation.

Dielectric waveguides like that shown in Fig. 3.5 are used in optical integrated circuits and are therefore likely to be of increasing importance.

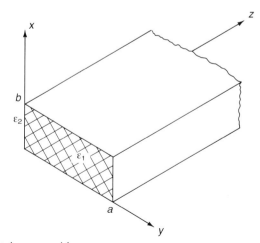

Fig. 3.5 Dielectric waveguide.

3.5 MONO-MODE AND MULTI-MODE OPTICAL FIBRES

Although dielectric guides are sometimes used at microwave frequencies
for special purposes the commonest use is in optical fibres. Figure 3.6
shows the three types of fibre which are in common use. In order these are
the monomode step-index fibre, the multimode step-index fibre and the
multimode graded-index fibre. These fibres are manufactured from glass,
silica or plastic with refractive indices in the range 1.0 to about 1.5. At the
present time they are normally used with carrier wavelengths in the range
0.82 to 0.85 μm. The theory of propagation in circular fibres is somewhat
difficult, but orders of magnitude for the quantities involved can be obtained
from the equations in the preceding section.

Consider first the monomode step-index fibre shown in Fig. 3.6(a). Let
us assume that the refractive index of the core is 1.5, that of the cladding is
1.4, and that the operating wavelength is 0.82 μm. From (3.22) the core
diameter at which this signal is cut off is 1.5 μm and the core diameter at
which the next higher mode will propagate at this wavelength is 3.0 μm. It
follows that the core diameter must lie somewhere between these limits if
only the lowest mode is to propagate. More exact calculations show that
the cut-off wavelength of the lowest mode for a circular fibre of diameter d
is given by

$$\lambda_c = \frac{\pi d}{2.405} \sqrt{(n_1^2 - n_2^2)}. \qquad (3.23)$$

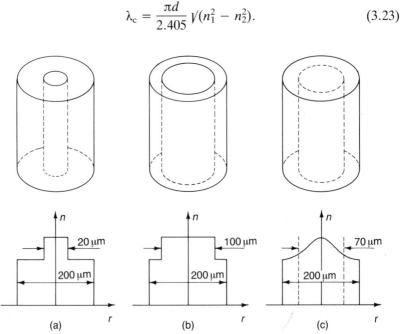

Fig. 3.6 Optical fibres: (a) mono-mode step-index, (b) multi-mode step-index and
(c) graded-index fibres.

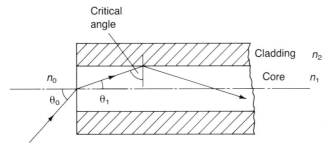

Fig. 3.7 Geometry of light acceptance by an optical fibre.

For the case given above $d = 1.17 \, \mu m$ showing that the results of the approximate calculation are not too far from the truth.

The other piece of information which is of interest is the angle of incidence of the wave at the cut-off wavelength. From (3.5) this is

$$\theta = \sin^{-1} \left(\frac{\varepsilon_2}{\varepsilon_1} \right)^{\frac{1}{2}} = \sin^{-1} \left(\frac{n_2}{n_1} \right) \tag{3.24}$$

giving a numerical value of 69° for the fibre described above. This angle is referred to as the critical angle of the fibre. Figure 3.7 illustrates the problem of launching a wave down a monomode step-index fibre. A ray in air is incident on the end of the fibre at an angle θ_0. On entering the core of the fibre it is refracted so that it makes an angle θ_1 with the axis. This angle must be less than 90° minus the critical angle. There is therefore a maximum value of the angle θ_0 for which the incident light will be totally reflected and so captured by the guide. This is known as the maximum acceptance angle θ_m and the cone which it defines is known as the acceptance cone. Another figure of merit which is sometimes used is the numerical aperture defined by

$$NA = \sin \theta_m. \tag{3.25}$$

It can be shown that

$$NA = n_1^2 - n_2^2. \tag{3.26}$$

The very small core diameter of a monomode step index fibre means that it is essential to use a laser as the signal source in order to couple sufficient power into the fibre. These fibres have low dispersion and a very wide usable bandwidth (up to 3 GHz km) but are very difficult to splice together. Their principal use is for submarine cables.

If the core of a step-index fibre is made larger then a number of different modes can propagate. Such a fibre is known as a multi-mode step-index fibre (Fig. 3.6(b)). The core diameter is typically one to two orders of magnitude larger than that of the monomode fibres discussed above. This makes the numerical aperture much larger so that sufficient light can be coupled into the fibre from a light emitting diode (LED). It also makes the splicing of the fibres possible. The disadvantage is that the bandwidth at up to 200 kHz km is much less than that of a monomode fibre. It is easy to see why this is so. Light entering the fibre is transmitted in a number of modes each having a different axial propagation constant. Thus a short pulse of light injected into the guide emerges as a longer pulse because of the different time delays of the different modes. Eventually the pulse dispersion can be so great as to prevent the correct reception of the data at the end of the fibre.

Multi-mode step-index fibres are the cheapest to manufacture. They are in common use for data links of all kinds especially short-distance ones since the bandwidth diminishes with the length of the fibre. For military and avionic systems they have the advantage that they are secure and immune to electromagnetic interference.

The graded-index fibre shown in Fig. 3.6(c) represents a compromise between the two type of fibre already discussed. The core diameter is relatively large so many modes can propagate. The waves are guided by refraction rather than by total internal reflection so the mathematics becomes rather involved. As the refractive index diminishes towards the outside of the fibre a ray making large excursions from the axis tends to spend more time in the low refractive index (high phase velocity) region. This compensates to some extent for the extra path length so that a graded-index fibre is less dispersive than a comparable step-index fibre. Typical bandwidths are from 200 MHz km to 3 GHz km, making this kind of fibre suitable for medium-distance telecommunication links.

3.6 RADOMES, WINDOWS AND OPTICAL BLOOMING

We have noted that there is a close analogy between the propagation of plane waves through different media and the propagation of TEM waves on transmission lines. This analogy enables us to use transmission-line methods for solving plane-wave problems. In particular the half- and

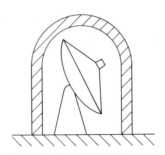

Fig. 3.8 Use of a radome to protect an antenna.

quarter-wave transformers have important practical applications. Figure 3.8 shows one such application. A radar antenna is enclosed by a dielectric enclosure called a radome to protect it from the weather. It is clearly desirable to design the radome in such a way that it does not reflect a lot of power back into the antenna. To investigate this problem we recall the equation for the transformation of impedances on lossless transmission lines

$$\frac{Z_{in}}{Z_0} = \frac{Z_L + jZ_0 \tan kl}{Z_0 + jZ_L \tan kl}. \tag{3.27}$$

This equation gives the input impedance when a line having characteristic impedance Z_0 and electrical length kl is terminated by an impedance Z_L. Making use of the analogy between plane waves and transmission lines we replace the characteristic impedance of the line by the wave impedance of the dielectric and the load impedance by the wave impedance of free space. If the radome is to be matched to the incident wave then the input impedance must also be the wave impedance of free space. The condition for this to occur is $\tan kl = 0$ that is $l = n\pi/2$, where $n = 1, 2, 3$, etc. Thus a sheet of dielectric which is an integral number of half wavelengths thick does not reflect any of the incident wave. Reference to (1.84) shows that this is because the waves reflected from the front and back faces of the sheet are in antiphase and so cancel each other out. In practice the problem is more difficult than this because the radome lies in the near field region of the antenna (see Ch. 5). For further information on radomes see Rudge et al. (1982–3).

Example

Find a suitable thickness for a perspex radome to operate at 10 GHz. ($\varepsilon_r = 2.6$ for perspex at 10 GHz).

Solution

At 10 GHz the free-space wavelength is 30 mm so the wavelength in the perspex is $30/\sqrt{2.6} = 18.6$ mm. Possible thicknesses for the radome are therefore 9.3 mm, 18.6 mm, 27.9 mm, etc. The thickness actually used would depend upon the need for adequate strength and rigidity.

The same principle can be used whenever it is necessary to put a dielectric barrier between two regions having the same wave impedance. Thus it can be used to provide pressure-tight windows in waveguides. The disadvantage of the technique is that it is narrow band because a small change of frequency causes the electrical length of the dielectric region to depart from $\lambda/2$.

When the thickness of the dielectric slab is not equal to an exact number

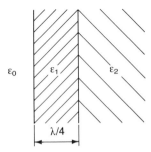

Fig. 3.9 Use of a quarter-wave coating to reduce the reflection coefficient at a dielectric boundary.

of half wavelengths it is still possible to reduce the reflected signal intensity by using the principle of the quarter-wave transformer to match the wave impedance in the dielectric to that of free space. Figure 3.9 shows how this is accomplished. The dielectric slab (permittivity ε_2) is coated with a thin layer of a dielectric of permittivity ε_1. The thickness of this layer must be an odd number of quarter wavelengths, the wavelength, of course, being calculated for the material of the layer. A common application of this technique is in the 'blooming' of lenses for cameras and other optical instruments. The impedances are only matched exactly at one wavelength, but the band of wavelengths in the visible spectrum is less than one octave so the mismatches at the red and blue ends of the spectrum are not too serious if the system is designed at the centre frequency. Blooming is achieved by vacuum deposition of materials such as magnesium fluoride ($n = 1.38$) and cryolite ($n = 1.36$) on the surface of the glass. These materials have refractive indexes which are rather higher than the ideal figure. It is nevertheless possible to reduce the reflection to less than 1% of the incident white light (Longhurst, 1973).

Example

A lens is made of crown glass (refractive index 1.52). Find the thickness and refractive index of a surface coating which will eliminate reflections from the surface of the glass at a wavelength of 0.5 μm.

Solution

Since, from (1.79), the wave impedance in a dielectric is inversely proportional to the square root of the relative permittivity it is also inversely proportional to the refractive index. The refractive index of the surface coating must be the geometric mean of the refractive indices of the glass and of free space, that is $\sqrt{1.52} = 1.23$.

The wavelength of the light in the surface layer is obtained by multiplying the free-space wavelength by the refractive index. Thus the thickness of the layer should be $(0.25 \times 0.5 \times 1.23) = 0.153\,\mu\text{m}$.

3.7 QUASI TEM WAVEGUIDES

A number of different types of two-wire TEM transmission lines were introduced in Chapter 2. Several other types of two-wire line are illustrated in Fig. 3.10. Superficially these are like the transmission lines shown in Fig. 2.1 but a little thought reveals that there are differences. The most obvious one is that the lines shown in Fig. 3.10 are not embedded in a homogeneous dielectric. Take the case of microstrip (Fig. 3.10(a)) as an example. The signal is guided by the strip conductor on the top of the dielectric substrate and the ground plane conductor beneath it. For purposes of analysis the region above the ground plane can be divided into two parts: the region within the substrate and the whole of the air space above it. In each of these regions the propagating waves must satisfy Maxwell's equations. If we suppose that this line carries a TEM wave like the lines discussed in Chapter 2 then the phase velocity of the parts of the wave within the two regions will differ from each other. In other words the wave in the air would travel faster than that in the dielectric and get out of step with it. If that were to happen the fields would no longer satisfy the boundary conditions at the interface between the two regions. We conclude, therefore, that a microstrip line cannot propagate a pure TEM wave. A similar argument can be applied to each of the other lines shown in Fig. 3.10.

Microstrip is much the commonest of the three types of line shown in Fig. 3.10. It is used extensively in the hybrid and monolithic integrated

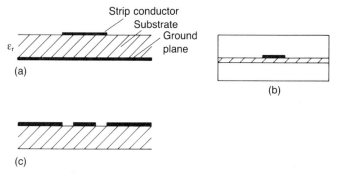

Fig. 3.10 Quasi-TEM transmission lines: (a) microstrip, (b) suspended substrate stripline and (c) coplanar waveguide.

circuits which are used for all low-power microwave signal processing. Suspended stripline has the advantage of low loss and it is used to make microwave filters whilst the balanced arrangement of coplanar waveguide is useful for the realization of balanced mixer circuits (Edwards, 1981). For all three it turns out that at low frequencies (below 2 GHz in the case of microstrip) the parameters of the line can be computed to a good approximation by static analysis. For this reason these lines can be refered to as 'quasi-TEM lines'. The discussion which follows will concentrate on microstrip.

As a starting point for the theory of microstrip consider the lines shown in Fig. 3.11. The wire over a ground plane shown in Fig. 3.11(a) can be analysed by elementary techniques (Carter, 1986) to give an expression for the capacitance per unit length

$$C = \frac{2\pi\varepsilon}{\ln{(4h/d)}} \qquad (3.28)$$

for $h \gg d$. Because this line is surrounded by a homogeneous dielectric it carries a TEM wave whose phase velocity is

$$v_p = \frac{1}{\sqrt{(\varepsilon\mu)}} = \frac{1}{\sqrt{(LC)}}, \qquad (3.29)$$

where C and L are the capacitance and inductance per unit length. The characteristic impedance is therefore

$$Z_0 = \sqrt{\left(\frac{L}{C}\right)} = \frac{1}{2\pi}\sqrt{\left(\frac{\mu}{\varepsilon}\right)}\ln\left(\frac{4h}{d}\right). \qquad (3.30)$$

The strip line over a ground plane shown in Fig. 3.11(b) is likewise surrounded by a uniform dielectric and therefore supports a TEM wave. This arrangement is not readily analysed by analytical methods but its resemblance to that of Fig. 3.11(a) suggests that it might be possible to find an empirical relationship between the dimensions of the two systems so that the strip conductor can be represented by an equivalent cylindrical wire. Such a formula has been suggested by Springfield (see Liao, 1980)

$$d = 0.67w\left(0.8 + \frac{t}{w}\right), \qquad (3.31)$$

where t/w is in the range 0.1 to 0.8.

Since the dielectric substrate of microstrip occupies only part of the space around the conductors it might be expected that the properties of the line would lie somewhere between those of an air-spaced line and one completely immersed in dielectric. This suggests that the properties of a microstrip line could be calculated by replacing the permittivity in (3.30) by an effective permittivity. A formula for this purpose has been suggested by Di Giacomo (1958)

Fig. 3.11 Dimensions for lines over ground planes: (a) thin wire and (b) strip conductors.

$$\varepsilon_{re} = 0.475\varepsilon_r + 0.67, \qquad (3.32)$$

where ε_r is the relative permittivity of the substrate material and ε_{re} is the effective relative permittivity to be used in (3.30). Combining (3.30), (3.31) and (3.32) gives an expression for the characteristic impedance of a microstrip line whose width is less than the substrate thickness

$$Z_0 = \frac{377}{2\pi} \frac{1}{\sqrt{(0.475\varepsilon_r + 0.67)}} \ln\left(\frac{5.97h}{0.8w + t}\right). \qquad (3.33)$$

This formula gives useful results up to a ratio w/h of 0.8.

For the extreme case of a line whose width is large compared with the thickness of the substrate it is possible to neglect the fringing fields so that the capacitance per unit length is just

$$C = \varepsilon w/h, \qquad (3.34)$$

which gives

$$Z_0 = \frac{377}{\sqrt{\varepsilon_r}} \frac{h}{w} \qquad (3.35)$$

for the characteristic impedance. Note that this time the effective permittivity is not used because of the assumption implicit in (3.34) that all the electric field between the conductors lies in the dielectric substrate. The relationship between the two formulae given for the characteristic impedance can be illustrated by considering an example.

Example

Estimate the variation of characteristic impedance with line width for microstrip having the following parameters

$$h = 0.5\,\text{mm}, \ t = 0.05\,\text{mm}, \ \varepsilon_r = 9.6 \ \text{(alumina)}.$$

Solution

For narrow linewidths we use (3.30), (3.31) and (3.32). From (3.32) the effective value of the relative permittivity is 5.23. The values of d and Z_0

may then be computed for the range of values of w for which this approximation is valid.

w (mm)	d (mm)	Z_0 (Ω)
0.1	0.087	82.1
0.2	0.141	69.5
0.3	0.194	61.1
0.4	0.248	54.7

For broad lines (3.35) is employed. The smallest value of w for which it is reasonable to suppose that the formula is valid is 2.0 mm. Computations for this and larger values of w yield the following.

w (mm)	Z_0 (Ω)
2.0	30.4
2.5	24.3
3.0	20.3

These figures are plotted in Fig. 3.12. The line has been sketched in so that it is asymptotic to the two curves as $w \rightarrow$ zero and $w \rightarrow$ infinity. From this it seems that the error in (3.33) is about 10% at worst for $w/h < 0.8$. Formula (3.35) is apparently valid to the same accuracy for $w/h > 3.0$. The curve may be expected to give values correct to within 10% in the region where neither formula is valid. Thus a 50 Ω line should have a width of around 0.8 mm which value is accurate enough for preliminary design calculations. The range of impedance values shown in this example is very typical of those normally used in microstrip circuits.

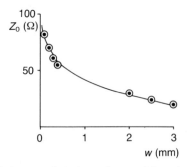

Fig. 3.12 Variation of characteristic impedance with strip width for a microstrip transmission line.

The preceding example has been included to show how the formulae can be used and what their limitations are. Methods of this kind involving effective and equivalent values are very commonly used to provide simple design methods for a whole range of complex problems where a full field solution would be too slow or difficult. An important application of these formulae is in computer-aided design (CAD) packages. A number of CAD packages are in common use for microstrip circuit design and these make use of formulae which are more complicated than those discussed above but which have been derived by very similar arguments. In particular these formulae make use of expressions for the effective permittivity which are frequency dependent to allow for the facts that the mode propagated is not a true TEM mode and that the line is therefore dispersive. Full details can be found in Edwards (1981) and Getsinger (1973).

Finally, it is important to remember that, like all other transmission lines, microstrip can propagate higher-order modes. These modes, which are cut off within the normal frequency band of the microstrip, set an upper limit on the frequency for which it can be used and also produce reactive parasitic effects at discontinuities in the line. This subject is discussed further in Chapter 6.

3.8 NON-TEM WAVEGUIDES

Besides the quasi TEM waveguides discussed in the previous section a number of other guides are sometimes employed in low power microwave circuits. These are shown in Fig. 3.13.

Slot line is the dual of an isolated strip line. It is useful when it is necessary to include shunt components in a circuit. The range of characteristic impedances which can be achieved is 60 to 200 Ω but the loss is rather high. An important use of slot line is in slot antennas. These are discussed in Chapter 5.

Fin line is superficially like the ridge waveguides discussed in Chapter 2. The range of impedances which can be realized is wide (10 to 400 Ω) and the losses are low.

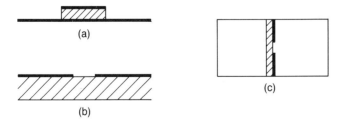

Fig. 3.13 Non-TEM waveguides: (a) image line, (b) slot line and (c) fin line.

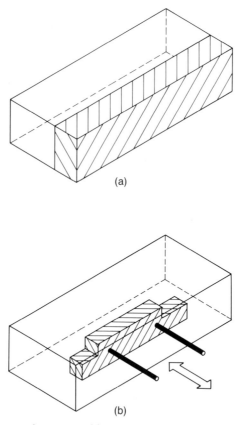

Fig. 3.14 (a) Rectangular waveguide partially filled with a dielectric material. (b) A waveguide moving vane phase changer.

Image line resembles dielectric waveguide. It can be made with very low loss and finds its applications at frequencies above 100 GHz.

The properties of rectangular and circular waveguides can be modified by the addition of dielectric materials. Thus, if a rectangular waveguide is filled with a dielectric material the free-space wavelength λ_0 in (2.35) is replaced by the wavelength of TEM waves in the dielectric. Dielectric-filled sections of waveguide are used as windows to separate regions of guide in which the gas pressures are different.

If a guide is partially filled with dielectric as shown in Fig. 3.14(a) then the guide wavelength will lie somewhere between that of an empty guide and one completely filled with dielectric. This principle is used to make waveguide phase-shifters as shown in Fig. 3.14(b). A dielectric vane is arranged so that it can be moved across the guide from the region of weak electric field near the wall to the strong field at the centre. The ends of the vane are stepped or tapered to ensure a satisfactory match. The effect of the vane is to modify the shunt capacitance of the guide (see Fig. 2.18) and, hence, its cut-off frequency (2.53). The device therefore acts as a variable phase shifter. The phase shift varies with frequency because the length of the vane in wavelengths is frequency dependent. An alternative form of phase shifter (Fox, 1947) uses a circular waveguide and a vane which rotates about the axis of the guide rather like the rotary vane attenuator described in Section 4.6.

3.9 CONCLUSION

In this chapter we have considered the reflection and refraction of waves at the boundaries between dielectric materials. The reflection of waves has been shown to depend upon their polarization, a phenomenon which is employed in the windows of gas lasers. Consideration of the circumstances in which waves can be trapped within dielectric strips and rods led on to a discussion of the properties of dielectric and optical fibre waveguides. The advantages of single-mode and multi-mode fibres were explored.

Consideration of the reflection of waves at multiple boundaries such as those encountered in radomes and microwave windows led to an understanding of the principles underlying the design of these components. In the optical field the same approach is used to minimize the reflections from the surfaces of lenses and other optical components.

Finally a number of waveguides incorporating inhomogeneous dielectrics were discussed. These guides, especially microstrip, are of considerable practical importance but are not easily analysed. The development of design formulae for microstrip was used to illustrate the kinds of methods which are commonly employed.

EXERCISES

3.1 Calculate the critical angle for total internal reflection for epoxy resin ($\varepsilon_r = 3.5$).

3.2 Calculate the relative permittivity of water at optical wavelengths given that the critical angle is 49°.

3.3 Calculate the Brewster angle for soda glass ($\varepsilon_r = 6.1$), epoxy resin ($\varepsilon_r = 3.5$) and perspex ($\varepsilon_r = 2.6$).

3.4 Calculate the cut-off frequencies for the lowest TE mode in a sheet of alumina ($\varepsilon_r = 8.9$) 1 mm thick. What would be the new cut-off frequency if the alumina were potted in epoxy resin ($\varepsilon_r = 3.5$)?

3.5 A radome made of glass-reinforced plastic having a relative permittivity of 4.5 is designed for use at 5 GHz. Calculate the minimum thickness which can be used and the reflection coefficients at 4.5 GHz and 5.5 GHz.

3.6 It is proposed that the broadband reflection coefficient of the radome in the previous question should be improved by coating it on both sides with perspex ($\varepsilon_r = 2.6$). Suggest a suitable thickness of perspex and calculate the new reflection coefficients at 4.5 and 5.5 GHz.

3.7 Calculate the dimensions of microstrip lines having conductors 0.05 mm thick and fused quartz substrate ($\varepsilon_r = 3.8$) which have characteristic impedances of 25, 50 and 100 Ω.

Waves with imperfectly conducting boundaries

4

4.1 WAVES INCIDENT NORMALLY ON A CONDUCTING SURFACE

In Chapter 2 the reflection of waves from perfectly conducting boundaries was discussed. In reality no boundaries (other than superconducting ones) are lossless so it is necessary to consider what the effects of imperfectly conducting boundaries are.

Consider the case of a plane wave in a lossless medium incident normally on a semi-infinite conducting slab as shown in Fig. 4.1. We will assume that the slab is a good conductor as defined in Section 1.5. Let the amplitudes of the incident, reflected and transmitted waves at the surface of the slab be E_i, E_r and E_t as shown. Then, from (1.84) and (1.85)

$$\frac{E_r}{E_i} = \frac{Z_2 - Z_1}{Z_2 + Z_1} \tag{4.1}$$

$$\frac{E_t}{E_i} = \frac{2Z_2}{Z_2 + Z_1}, \tag{4.2}$$

where the wave impedances of the two materials are

$$Z_1 = \sqrt{\left(\frac{\mu_1}{\varepsilon_1}\right)} \quad \text{and} \quad Z_2 = \frac{(1 + j)}{\sigma\delta} \tag{4.3}$$

from (1.29) and (1.45). Note that, because Z_2 is complex with a phase angle of 45°, the transmitted wave is out of phase with the incident wave.

The power absorbed by the surface can be calculated by a simple application of Poynting's theorem. The transmitted wave travels into the slab and is completely absorbed by it. The power absorbed per unit area is therefore equal to the power flowing into unit surface area of the slab which is just the magnitude of the Poynting vector. Thus, from (1.46)

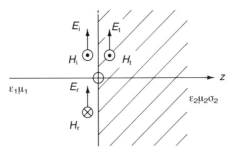

Fig. 4.1 Incident, reflected and transmitted waves for electromagnetic waves incident normally on the surface of a material having finite conductivity.

$$S_t = \tfrac{1}{4}\sigma\delta E_t^{\,2} \tag{4.4}$$

$$S_t = \tfrac{1}{4}\sigma\delta \left| \frac{2Z_2}{Z_1 + Z_2} \right|^2 E_i^{\,2}$$

$$= \tfrac{1}{2}\sigma\delta Z_1 \left| \frac{2Z_2}{Z_1 + Z_2} \right|^2 P_i. \tag{4.5}$$

From this equation it is evident that the units of $\sigma\delta$ must be Ω^{-1} (siemens). It is convenient to define the *surface resistance* (or sheet resistivity) of the slab by

$$R_s = 1/\sigma\delta. \tag{4.6}$$

The wave impedance of the material can then be written

$$Z_2 = R_s(1 + j). \tag{4.7}$$

From (4.5) the transmission loss at the interface is

$$L = -10 \log_{10} \left(\frac{1}{2} \frac{Z_1}{R_s} \left| \frac{2Z_2}{Z_1 + Z_2} \right|^2 \right). \tag{4.8}$$

Very often $|Z_1| \gg |Z_2|$ so that, to a good approximation,

$$L = -10 \log_{10} \left[\frac{2|R_s(1 + j)|^2}{R_s Z_1} \right]$$

$$= -10 \log_{10} \left(\frac{4R_s}{Z_1} \right). \tag{4.9}$$

The concept of surface resistance can be given a simple physical interpretation. First we recall that, as shown in (1.44), a wave propagating through a good conductor decays exponentially with a decay constant equal to the skin depth δ. We recall also that, for a good conductor, δ is typically of the order of a millimetre or less. Consider therefore the approximation that

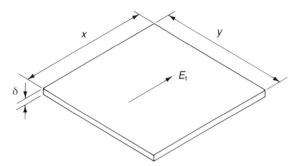

Fig. 4.2 Geometry of a thin sheet of resistive material.

the power dissipated in the material is dissipated uniformly in a surface layer of thickness δ as shown in Fig. 4.2. The resistance of the sheet measured in the x direction is

$$R = \frac{x}{\sigma \delta y} = \frac{x}{y} R_s. \tag{4.10}$$

For the special case of a square sheet it can be seen that $R = R_s$ regardless of the size of the square. For this reason surface resistance is usually written as so many 'ohms per square' sometimes abbreviated to $\Omega/\square$.

If all the current in the conductor is considered to flow in the skin depth then the boundary condition on the magnetic field is

$$H = 2H_i = A \tag{4.11}$$

where A is the current flowing in unit width of the surface. The almost complete reflection of the incident wave at the surface together with the accompanying reversal of the electric field vector means that the component of the magnetic field parallel to the surface is the sum of the tangential components of the incident and reflected magnetic field vectors (i.e. twice the incident field). The power dissipated per unit area is thus

$$P = \tfrac{1}{2}A^2 R_s = \tfrac{1}{2}H^2 R_s = 2H_i^2 R_s. \tag{4.12}$$

Now from (4.2) making the approximation that $Z_2 \ll Z_1$

$$E_t = 2E_i R_s(1 + j)/Z_1$$

so
$$E_t^2 = 8R_s^2 E_i^2/Z_1^2$$
$$= 8R_s^2 H_i^2. \tag{4.13}$$

Substituting this in (4.4) gives (4.12) showing that the two approaches are consistent with each other.

It is frequently helpful to use this way of treating the absorption of power in a conducting surface. The use of the magnetic field and the accompanying surface current density is usually easier than the use of the electric field though both methods can be shown to give the same answers.

We may therefore think of all the power being dissipated in a surface layer, having a surface resistance R_s, whose thickness is equal to the skin depth. This approximation is not as crude as it seems. The power flow varies as the square of the electric field strength and therefore decays as $\exp -z/2\delta$ as the wave penetrates into the slab. At the skin depth the power is only 13.5% of that at the surface so that 86.5% of the power entering the slab is absorbed within the skin depth.

Example

Estimate the difference between the intensity of a 1 kHz radio wave received by a submarine on the surface of the sea and that received when the submarine is submerged to a depth of 10 m.

Solution

Because sea water is an electrical conductor it is not possible for radio waves to penetrate very far into it. This makes radio communication with submerged submarines very difficult. To make the skin depth as large as possible a very low carrier frequency must be used and that sets severe limitations on the possible rate of transmission of data.

The relative permitivity of sea water is about 81 and its conductivity about 4 S. Substituting these figures together with the given frequency into the right hand side of (1.37) yields

$$(\delta + j\omega\varepsilon) = (4 + 4.5 \times 10^{-6} \, j) \qquad (4.14)$$

so that we are justified in treating sea water as a good conductor in this problem. The skin depth is, from (1.43),

$$\delta = \sqrt{(2/\omega\sigma\mu)} = 8.0 \, \text{m}. \qquad (4.15)$$

It is not sufficient merely to compute the decay of the signal as it passes through the water. We must also allow for the reflection of some of the incident power at the air–water interface. To simplify the calculation we will assume that the wave is incident normally on the sea surface. The wave impedances are

$$Z_1 = 377 \, \Omega \qquad \text{and} \qquad Z_2 = 0.031(1 + j) \, \Omega. \qquad (4.16)$$

The very large difference in the magnitudes of the two impedances allows us to use the approximate expression for the transmission loss given in (4.9). The transmission loss is therefore

$$L_t = -10 \log_{10} \left[\frac{4 \times 0.031}{377} \right] = 35 \, \text{dB}. \qquad (4.17)$$

The signal power under the sea is proportional to the square of the electric field amplitude. At a depth of 10 m the power is

$$P = P_t \exp - \left[\frac{2 \times 10}{\delta} \right]$$

$$= 0.082 P_t \tag{4.18}$$

so that the signal transmission loss is

$$L = -10 \log_{10} \left(\frac{P}{P_t} \right) = 11 \, \text{dB}. \tag{4.19}$$

Combining the two transmission losses shows that the difference between the signal received on the surface and that received at a depth of 10 m is 46 dB.

4.2 TRANSMISSION THROUGH A THIN CONDUCTING SHEET

A very important practical case arises when the conducting material is in the form of a thin sheet rather than a thick slab as assumed in the previous section. 'Thin' in this context means 'having a thickness of the same order of magnitude as the skin depth'.

Figure 4.3 shows the arrangement of the problem. Because it is likely that some of the forward wave in the conductor will be reflected at B it is necessary to include this possibility in the calculations. The amplitudes of the forward and backward waves in each region are shown in the diagram.

If the thickness of the conductor is equal to the skin depth then the transmission loss is, from the previous section

$$L = -10 \log (0.135) = 8.7 \, \text{dB}. \tag{4.20}$$

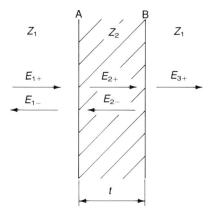

Fig. 4.3 Forward and backward waves for reflection by and transmission through a conducting sheet of finite thickness.

Thus, even if the forward wave is completely reflected at B and the backward wave completely reflected at A the doubly reflected signal has an amplitude which is 17.4 dB below that of the initial forward wave. If the thickness of the sheet is greater than the skin depth the effect of multiple reflections can be neglected. With this approximation the problem is reduced to the computation of the transmission losses at A and B and the transmission loss for the path A–B.

The transmission loss at A is given by (4.9) and the transmission loss through the sheet is

$$L_{AB} = -10 \log_{10} [\exp - (2t/\delta)]. \tag{4.21}$$

At B we have, by analogy with (4.2),

$$\frac{E_{3+}}{E_{2+}} = \frac{2Z_1}{Z_1 + Z_2} \simeq 2 \tag{4.22}$$

if $|Z_1| \gg |Z_2|$. This apparently surprising result occurs because the boundary condition at B is very nearly an open circuit. Very nearly all of the incident wave is reflected with a phase change of 180° making the electric field at the boundary twice that of the incident wave. The power in the transmitted wave is actually very small compared with the incident powder despite the magnitudes of the electric fields because of the differences in the impedances of the two media.

The incident power is

$$P_{2+} = |E_{2+}|^2/4R_s \tag{4.23}$$

from (1.46) and the transmitted power is

$$P_{3+} = |E_{3+}|^2/2Z_1$$
$$= 2|E_{2+}|^2/Z_1. \tag{4.24}$$

The transmission loss at B is therefore

$$L_B = -10 \log_{10} \left(\frac{P_{3+}}{P_{2+}}\right)$$

$$= -10 \log_{10} \left(\frac{8R_s}{Z_1}\right). \tag{4.25}$$

The significance of these results can best be illustrated by an example.

Example

Calculate the attenuation of a 1 MHz radio wave by a sheet of aluminium 0.2 mm thick.

Solution

The conductivity of aluminium is $3.5 \times 10^7 \, \text{S m}^{-1}$ so the skin depth at 1 MHz is 0.085 mm (from (1.44)) and the surface resistance is $0.34 \times 10^{-3} \, \Omega/\square$. These figures show that we are justified in neglecting multiple reflections within the aluminium and in assuming that the wave impedance in the aluminium is much less than that in the air. The transmission losses are, from (4.9), (4.21) and (4.25),

$$L_A = 54 \, \text{dB}$$
$$L_{AB} = 20 \, \text{dB}$$
$$L_B = 51 \, \text{dB},$$

giving a total transmission loss of 125 dB.

4.3 ELECTROMAGNETIC SCREENING

In the example discussed in the previous section we saw how electromagnetic waves are attenuated by the presence of a conducting barrier. This leads naturally to a discussion of screening against electromagnetic interference. This topic is part of the subject known as electromagnetic compatibility (EMC). EMC is concerned with all the possible ways in which electronic equipment can be a source of interference or be susceptible to it. The steady increase in the number of possible sources of interference and the progress towards ever tighter packing densities in electronic equipment both make an understanding of the principles of EMC important to electronic engineers.

There is much more to the subject than finding ways of suppressing radio interference from cars and electric motors. In a modern aircraft, for example, a lot of electronic equipment is packed into an extremely small space. It is quite possible for signals radiated by a radar set to interfere with, say, the navigation system. If the interference is bad enough it could result in the loss of the aircraft. In addition to signals generated locally there is also the possibility of interference from high-power transmitters, from lightning and from solar storms. Perhaps the most dramatic example of all is the electromagnetic pulse (EMP) produced by a thermonuclear explosion. This pulse is powerful enough to destroy semiconductor devices in unprotected electronic equipment many miles from the explosion (Keiser, 1983).

In the previous section we saw that the screening effect of a conducting sheet can usually be divided into three parts, namely the reflection from the front of the sheet, the attenuation through it and the reflection from the back of the sheet. In addition, when the transmission loss through the sheet is small (typically less than 15 dB), multiple reflections can be important. The overall screening effectiveness can therefore be written

$$S = R + A + B, \tag{4.26}$$

where R is the transmission loss produced by both reflections, A is the attenuation loss for waves passing once through the sheet and B represents the effects of multiple reflections, all expressed in decibels. Note that B can be either positive or negative depending upon whether the multiply reflected waves interfere with each other constructively or destructively.

The reflection loss is, from (4.9) and (4.25),

$$R = -10 \log_{10} (4R_s/Z_1) - 10 \log_{10} (8R_s/Z_1). \tag{4.27}$$

Substituting for R_s and Z_1 we obtain, after a little manipulation,

$$R = 31.5 - 10 \log_{10} \left(\frac{f\mu}{\sigma}\right) dB. \tag{4.28}$$

The effects of reflections therefore decrease with increases in the frequency and of the permeability of the screen and increase with increases in conductivity.

The attenuation loss given by (4.21) may be written

$$A = -\frac{t}{\delta} 10 \log_{10} (e^{-2}) \tag{4.29}$$

which becomes

$$A = 15.4\sqrt{(f\sigma\mu)}t \, dB, \tag{4.30}$$

showing that the attenuation loss increases with the thickness of the sheet and also with the frequency, permeability and conductivity. This term changes faster than the logarithm in (4.28) so the overall screening effectiveness increases roughly as the attenuation loss.

If the sheet is very thin then the attenuation is negligible and all the screening is caused by the reflection loss. Such very thin screens can be made by evaporating metallic films on to dielectric surfaces. Multiple reflections cause the screening effectiveness to vary with frequency and it is a maximum when the film is a quarter wavelength thick. The reason for this can be seen by studying (4.1). The wave impedance of the film is much less than that of the surrounding space. This produces near short-circuit conditions at the front surface of the film and near open-circuit conditions at the back. Equation (4.1) shows that there is a phase reversal at the first reflection but not at the second. Thus if the film is a quarter wavelength thick the reflections from the two surfaces are in phase and the total reflection is a maximum.

The theory of screening by a conducting sheet described above involves a number of simplifying assumptions which need to be examined. The first of these is that the incident signal is a plane wave. This is only true if the screen is at least several wavelengths from the source. Sometimes, especially at low frequencies, this will not be the case. In the limit when the frequency is zero we know that an electric field can be completely excluded from a perfectly conducting closed box (a Faraday cage) (Carter, 1986). Such a box is, however, completely transparent to a static magnetic field. Similarly, a closed box of a high-permeability material such as mumetal can act as an effective screen against a static magnetic field but is less effective against electric fields because of the relatively poor conductivity of mumetal. This suggests that when the screen is close to the source its screening effectiveness will depend upon whether the source is an electric or a magnetic one.

As the frequency rises the eddy currents induced in a conducting screen tend to exclude the magnetic field from it so increasing its screening effectiveness. Conversely the currents produced by a changing electric field act as sources of electric field on the further side of the screen so reducing its effectiveness as the frequency increases. Thus it is useful to discuss separately the electric and magnetic screening effectiveness of an enclosure. Figure 4.4 shows their behaviour for a typical enclosure.

The next approximation which needs discussion is the application of the solution for the normal incidence of plane waves on an infinite sheet to practical cases involving finite enclosures of irregular shapes. The exact solution of practical problems is very difficult and can only be achieved in simple cases by numerical methods (Akhtarzad and Johns, 1975). Generally, though, we do not require very high accuracy in the estimation of the effectiveness of a screen. A useful approximation is to consider the screening effectiveness of a thin spherical shell of radius R and thickness t. The general formulae given by Field (1983) are

$$S_M = 20 \log_{10} \left[\cosh \gamma t + \frac{1}{3} \left[\frac{\gamma R}{\mu_r} + \frac{2(\mu_r - 1)}{\gamma R} \right] \sinh \gamma t \right], \qquad (4.31)$$

where $\gamma = \sqrt{(j\omega\sigma\mu)}$, and

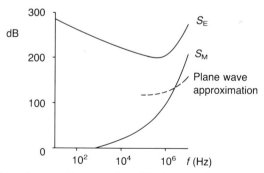

Fig. 4.4 Electric and magnetic screening effectiveness for an aluminium box as a function of frequency.

$$S_E = 20 \log_{10} \left[\frac{2\gamma \sinh \gamma d}{3\omega^2 \varepsilon_0 \mu R} \right]. \tag{4.32}$$

These expressions are a little involved. Useful approximations are:

1. the magnetostatic case

$$S_M = 20 \log_{10} \left[1 + \frac{2(\mu_r - 1)t}{3R} \right]; \tag{4.33}$$

2. the low-frequency approximations ($t \ll \delta$)

$$S_M = 10 \log_{10} \left[1 + \omega^2 \left(\frac{\mu_0 \sigma R t}{3} \right)^2 \right] \tag{4.34}$$

$$S_E = 20 \log_{10} \left[\frac{2\sigma t}{3\omega \varepsilon_0 R} \right]; \text{ and} \tag{4.35}$$

3. the high-frequency approximations ($t \gg \delta$)

$$S_M = 20 \log_{10} \left[\frac{R}{3\sqrt{2}\mu_r \delta} e^{t/\delta} \right] \tag{4.36}$$

$$S_E = 20 \log_{10} \left[\frac{\sigma \delta}{3\sqrt{2}\omega \varepsilon_0 R} e^{t/\delta} \right]. \tag{4.37}$$

To see what these formulae imply let us consider an example.

Example

Estimate the screening effectiveness of a rectangular aluminium box 0.2 mm thick whose dimensions are 50 mm × 100 mm × 200 mm over the frequency range 10 Hz to 10 MHz.

Solution

Field (1983) suggests that the screening effectiveness of a rectangular box can be estimated by calculating it for a spherical shell having the same volume. Thus $R = 62$ mm. The skin depth is equal to the shell thickness at 36 kHz so we can use (4.34) and (4.35) from 10 Hz to 10 kHz and (4.36) and (4.37) from 100 kHz to 10 MHz. The results of these calculations are shown in Table 4.1.

These figures suggest that the screening effectiveness of 125 dB calculated using the plane-wave approximation is an over-estimate if the source of interference is a magnetic source. It is evident that the figures in the table follow the pattern shown in Fig. 4.4.

Table 4.1

f	δ (mm)	S_M (dB)	S_E (dB)
10 Hz	12	0	283
100 Hz	3.8	0.06	263
1 kHz	1.2	3.6	243
10 kHz	0.38	21.2	223
100 kHz	0.12	56	204
1 MHz	0.038	97	205
10 MHz	0.012	206	274

Example

What is the effect of replacing the aluminium enclosure of the previous example with one made of mumetal having the same dimensions?

Solution

Mumetal (a special magnetic screening alloy) has a relative permeability of 80×10^3 and a conductivity of $1.74 \times 10^6 \, S \, m^{-1}$. The shell thickness is equal to the skin depth at 45 Hz. This time it is possible to calculate the magnetic screening effectiveness under d.c. conditions using (4.33). Table 4.2 shows the results of the calculations.

Table 4.2

f	δ (mm)	S_M (dB)	S_E (dB)
0	–	65	–
10 Hz	0.42	56	277
100 Hz	0.13	34	317
1 kHz	0.042	72	315
10 kHz	0.013	175	648

These figures again show the same pattern as Fig. 4.4 but this time the magnetic screening is much better at low frequencies. The reduction in the electric screening effectiveness at low frequencies is because the impedance mismatch at the surface of the mumetal is less than that at the surface of the aluminium.

The values of screening effectiveness calculated in the preceding examples are so high that it appears that effective screening of electronic equipment should present no problems. The fact that very real problems do exist is a consequence of the other assumptions which have been made in the calculations. These are:

1. the enclosure is not resonant at any frequency within the range of interest; and
2. the enclosure is perfect with no hole in it of any kind.

Assumption 1 is equivalent to saying that the enclosure is small compared with the wavelength of the waves within it at all frequencies which matter. This is not necessarily the case when the frequencies are in the gigahertz region (including any signal harmonics), or when the circuits are mounted on high-permittivity substrates (e.g. alumina) or encapsulated in epoxy resin. We shall return to this point when resonant cavities are discussed in Chapter 6.

Assumption 2 is almost impossible to satisfy in any real system. Since a circuit must do something it must exchange energy or information with the outside world. This can only be achieved by making holes in the screen to allow wires or optical fibres to pass through. We shall see in the next chapter that the effect of such holes is a dramatic reduction in screening effectiveness. It follows that very great care is necessary in the design and construction of enclosures if they are not to be degraded in this way.

The final assumption which needs to be explored is that the enclosure is made wholly out of metal sheet. In many cases it is inconvenient to use sheet and other possibilities exist such as conducting plastics and metals in woven, braided or expanded mesh forms. It is to be expected that these materials will be less effective as screens than sheet metal but they can still provide useful screening. For a fuller discussion of the subject consult Field (1983) and Keiser (1979).

4.4 WAVES INCIDENT OBLIQUELY ON A CONDUCTING SURFACE

So far we have only considered the case where a wave is incident normally on a conducting surface. We now turn to the general case.

The wave impedance for waves incident obliquely on a surface with E parallel to the surface is

$$Z = Z_1/\cos \theta \tag{4.38}$$

from (1.97). Thus

$$\frac{E_r}{E_i} = \frac{Z_2 \cos \theta_1 - Z_1 \cos \theta_2}{Z_2 \cos \theta_1 + Z_1 \cos \theta_2} \tag{4.39}$$

and
$$\frac{E_t}{E_i} = \frac{2Z_2 \cos \theta_1}{Z_2 \cos \theta_1 + Z_1 \cos \theta_2} \qquad (4.40)$$

from (4.1) and (4.2) with Z_1 and Z_2 given by (4.3). The refraction of the waves at the boundary must obey Snell's Law (1.91) so

$$\frac{\sin \theta_2}{\sin \theta_1} = k_1 \delta. \qquad (4.41)$$

In any good conductor the skin depth is much less than the free-space wavelength so the right-hand side of (4.41) is normally very small. It follows that θ_2 is very close to zero so that the transmitted wave can be regarded as travelling normal to the boundary. Making this approximation and substituting for Z_2 gives

$$\begin{aligned} \frac{E_r}{E_i} &= \frac{(1 + j)R_s \cos \theta_1 - Z_1}{(1 + j)R_s \cos \theta_1 + Z_1} \\ &\simeq -1, \end{aligned} \qquad (4.42)$$

since Z_1 is normally much greater than R_s. The wave is, therefore, almost completely reflected with a phase reversal just as in the case of a perfect conductor. The transmitted wave amplitude is

$$\begin{aligned} \frac{E_t}{E_i} &= \frac{2(1 + j)R_s \cos \theta_1}{(1 + j)R_s \cos \theta_1 + Z_1} \\ &\simeq \frac{2(1 + j)R_s}{Z_1} \cos \theta_1, \end{aligned} \qquad (4.43)$$

showing that the power absorbed decreases as the angle of incidence increases.

When the incident wave has its magnetic field parallel to the boundary the wave impedances are given by (1.94) and we obtain the following expressions for the reflected and transmitted wave amplitudes with the same assumptions as before:

$$\frac{E_r}{E_i} = \frac{(1 + j)R_s - Z_1 \cos \theta_1}{(1 + j)R_s + Z_1 \cos \theta_1} \qquad (4.44)$$

and
$$\frac{E_t}{E_i} = \frac{2(1 + j)R_s}{(1 + j)R_s + Z_1 \cos \theta_1}. \qquad (4.45)$$

For most angles of incidence the wave impedance of medium 1 is much greater than that of the conductor and the wave is almost completely reflected with a phase reversal as before. However, when there is grazing incidence the two terms on the top line of (4.44) are of comparable magnitudes and an appreciable part of the incident power may be absorbed. The condition for mimimum reflection is

$$Z_1 \cos \theta_1 = R_s \qquad (4.46)$$

and then

$$\frac{E_t}{E_i} = \frac{j}{2 + j} \qquad (4.47)$$

and

$$\frac{E_r}{E_i} = \frac{2(1 + j)}{2 + j}. \qquad (4.48)$$

Note carefully that it is never possible for all the power to be absorbed because of the phase difference between the impedances of the two media. When condition (4.46) is satisfied the incident and reflected waves are no longer in antiphase. To get a feel for the numbers involved consider the case of a 1 MHz wave in air incident on an aluminium sheet. The surface resistance of aluminium is $0.34 \times 10^{-3} \, \Omega$ so the angle of incidence for minimum reflection would differ from 90° by only a few millionths of a degree.

4.5 LOSSES IN TRANSMISSION LINES AND WAVEGUIDES

The theory of the preceding section finds its application in the calculation of losses in transmission lines and waveguides. In a TEM transmission line the electric field is normal to the surface of the conductors and the magnetic field is tangential. The loss per unit area of the conductors is given by (4.12).

Example

Find the attenuation per unit length at 10 GHz of a 50 Ω semi-rigid coaxial cable whose centre conductor is 1 mm in diameter. The conductors are copper and the dielectric is polythene.

Solution

The characteristic impedance of a coaxial cable is given by

$$Z_0 = \frac{1}{2\pi} \sqrt{\left(\frac{\mu}{\varepsilon}\right)} \ln \left(\frac{b}{a}\right), \qquad (4.49)$$

where b and a are the radii of the outer and inner conductors, respectively (see Carter, 1986, p. 120). The relative permittivity of polythene is 2.25 so for a 50 Ω cable the inside diameter of the outer conductor must be 3.5 mm. The amplitude of the magnetic field is given by

$$H = \frac{I}{2\pi r}. \qquad (4.50)$$

The power absorbed per unit length of the outer conductor is

$$P_b = \frac{1}{2}\left(\frac{I}{2\pi b}\right)^2 R_s\, 2\pi b, \tag{4.51}$$

from (4.12) and (4.50). Similarly, the power absorbed per unit length by the inner conductor is

$$P_a = \frac{1}{2}\left(\frac{I}{2\pi a}\right)^2 R_s\, 2\pi a. \tag{4.52}$$

Thus the total loss per metre in the conductors is

$$P_L = \frac{I^2 R_s}{4\pi}\left(\frac{1}{a} + \frac{1}{b}\right), \tag{4.53}$$

so that the effective series resistance of the line is

$$R = \frac{R_s}{2\pi}\left(\frac{1}{a} + \frac{1}{b}\right). \tag{4.54}$$

For a low-loss line the attenuation constant $\alpha = R/2Z_0$ (see Appendix A). So the attenuation per metre is

$$L = 20\, \log_{10}\left[\exp - (R/2Z_0)\right]. \tag{4.55}$$

Now the conductivity of copper is $5.8 \times 10^7\, \mathrm{S\,m^{-1}}$ so the skin depth at 10 GHz is $0.66\,\mu m$ and the surface resistance is $0.026\,\Omega/\square$. The loss in the line is found, by substituting these figures into (4.55), to be $0.46\,\mathrm{dB\,m^{-1}}$. The loss in real lines is about three times this figure because the skin depth is much less than the surface roughness.

Another possible source of loss in the line is dielectric loss. The loss tangent of polythene at microwave frequencies is 0.0003. The dielectric loss per unit length is therefore from (1.52)

$$P_D = 20\, \log_{10}\left\{\exp - [\tfrac{1}{2}\omega\sqrt{(\varepsilon'\mu)}\delta]\right\} \tag{4.56}$$

(where δ is the loss tangent) and, substituting the numbers, we find that the dielectric loss is $0.2\,\mathrm{dB\,m^{-1}}$.

Provided that the losses are small it is justifiable to treat them separately and add the results. They can be modelled by a series resistance in the equivalent circuit of the line.

The attenuation of signals by other waveguiding systems can, in principle, be calculated in the same way. The non-uniform distribution of the currents in the conductors may make this difficult in practice. For further information on transmission line losses see Ramo *et al.* (1965) for waveguides and Edwards (1981) for microstrip.

The theory of wave propagation on lossy transmission lines is beyond the scope of this book. A useful treatment is given by Collin (1966). Fortunately it is usually possible either to neglect transmission-line losses or to treat them as lumped at a single point on the line. It should be noted that the characteristic impedance of a lossy line is complex and frequency dependent.

4.6 MICROWAVE ATTENUATORS

In the previous section we considered the effects of conductor and dielectric losses on the propagation of waves on transmission lines. These losses are generally small and can often be neglected. Sometimes, however, it is necessary to introduce much larger losses. Two common examples are attenuators (used to adjust signal levels) and loads (used to provide matched terminations). Microwave loads are discussed in the next section.

Consider first the realization of a fixed attenuator in a two-wire line. Superficially it would seem that all that is needed is to insert a lossy section into one of the conductors. Figure 4.5(a) shows this arrangement in coaxial line. The equivalent circuit of Fig. 4.5(b) shows the problem with this approach. The characteristic impedance of the lossy section is different from the rest of the line so there are mismatches at both ends of it. The reflections from these mismatches would beat with each other as the frequency changed, giving a variable transmission of power to the load. The alternative possibility of putting a thin resistive disc across the line shown in Fig. 4.5(c) has the equivalent circuit Fig. 4.5(d). This also introduces a mismatch into the line.

The solution is to use a combination of series and shunt elements as shown in Fig. 4.6. Figure 4.6(a) shows a combination of a lossy central conductor and a lossy disc in a coaxial line. Figure 4.6(c) shows a stripline attenuator having one series element and two shunt elements. The corre-

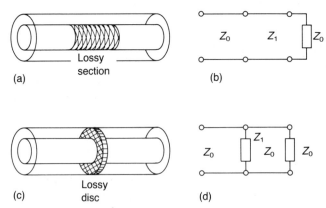

Fig. 4.5 Simple coaxial line attenuators and their equivalent circuits: series resistance (a) and (b), and shunt resistance (c) and (d).

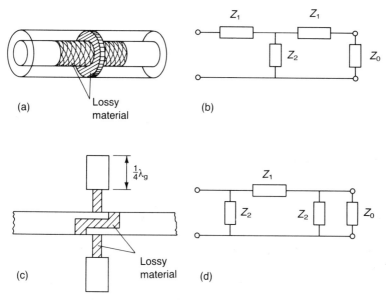

Fig. 4.6 Attenuators having both series and shunt elements with their equivalent circuits: coaxial line (a) and (b), and microstrip (c) and (d).

sponding equivalent circuits are shown in Fig. 4.6(b) and (d). By a suitable choice of the component values both these circuits can be matched to the transmission line. Provided that the physical sizes of the lossy elements are small (less than one eighth of a wavelength at the highest frequency) they can be regarded as lumped components. In that case the match depends only on the component values and not on their dimensions and it is possible to make an attenuator which has a good match and constant attenuation over a very wide frequency band. Coaxial attenuators are commercially available with attenuation flat to within ± 1 dB from d.c. to 18 GHz.

An interesting feature of the stripline design shown in Fig. 4.6(c) is the use of open-circuited quarter-wave sections of line to provide short-circuit terminations for the shunt resistors. This technique obviously limits the frequency band over which the attenuator will work correctly. The alternative would be to connect shorting wires through holes drilled in the substrate; this technique requires skilled manual operations.

Where variable coaxial attenuators are required two approaches are in common use. One method is to use a set of fixed attenuators with mechanical or PIN switches to select them. The other, the piston attenuator, employs the attenuation of a cut-off waveguide.

In waveguides a rather different approach is used to construct attenuators. Figure 4.7 shows two common arrangements. In both cases a resistive vane

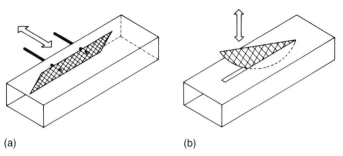

(a)   (b)

Fig. 4.7 Waveguide variable attenuators.

is arranged along the guide so that it is parallel to the electric field of the TE mode. The vane is typically around three wavelengths long. It is made either of a resistive card or of an insulator such as glass with a thin resistive film deposited on it. The section of guide with the vane in it is mismatched to the rest of the guide so the ends of the vane are tapered to provide a gradual transition. A match of around 1.05 VSWR can be achieved with a taper one wavelength long. Both the attenuators shown in Fig. 4.7 are variable. In Fig. 4.7(a) the vane is moved across the guide from the region of weak electric field at the side of the guide to the region of strong field at its mid-plane. In Fig. 4.7(b) the vane is lowered through a slot in the centre of the broad wall. Because the electrical length of the vanes varies with frequency the attenuation varies with frequency by a few decibels over an octave frequency band. The attenuation can be varied from zero to around 40 dB. One interesting variant of Fig. 4.7(b) has a dielectric vane with a neoprene tube fastened along its curved edge. Water is passed through the tube as a microwave absorber so making a high-power attenuator. Another type of high-power attenuator is described in Chapter 5.

The attenuators described in the previous paragraph are useful for level setting especially over narrow frequency bands. They have the advantage of being simple and relatively cheap. They can be calibrated if necessary but are not really suitable for use as attenuation standards for measuring purposes. Figure 4.8 shows a rotary vane attenuator. This type has very little variation of attenuation with frequency and its attenuation can be calculated so making it suitable for use as a reference. The device is symmetrical about its centre. The incoming rectangular waveguide A is connected via a transition B to a section of cylindrical waveguide C. B contains a fixed resistive vane arranged parallel to the broad wall of A. The purpose of this vane is to absorb any modes having horizontal electric field components so that the wave travelling from B into C is a pure TE_{11} mode polarized with its electric field vertical.

Section C contains a centrally placed attenuating vane and is capable of being rotated about its axis. The field of the incoming wave can be resolved

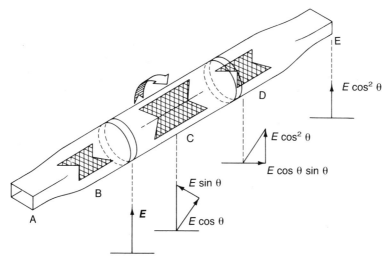

Fig. 4.8 Waveguide rotary vane attenuator.

into components $E \cos \theta$ normal to the plane of the vane and $E \sin \theta$ parallel to it. The vane is designed to provide very high attenuation for the component parallel to it. Thus the wave at the exit from C is polarized in the plane normal to the vane and has an amplitude $E \cos \theta$.

Section D is a transition from cylindrical to rectangular waveguide containing a fixed horizontal attenuating vane. The incoming wave can be resolved into a vertical component with amplitude $E \cos^2 \theta$ and a horizontal component with amplitude $E \cos \theta \sin \theta$. The horizontal component is completely absorbed by the vane. Thus the amplitude of the wave transmitted to the output rectangular waveguide E is $E \cos^2 \theta$. The attenuation is therefore

$$A = -40 \log_{10} (\cos \theta). \tag{4.57}$$

This depends only upon the angle of the vane in C so it is independent of frequency. These attenuators are usually calibrated to be read directly.

Compared with the moving-vane attenuators described earlier rotary-vane attenuators are bulky and expensive. They generally have excellent matches to the input and output guides because of the use of tapered ends to the vanes and tapered rectangular to circular waveguide transitions.

4.7 MICROWAVE LOADS

It is often important to have some means of providing a matched termination for a transmission line or waveguide. This is especially true in microwave measuring systems where any reflected signal will produce an error in the measurement.

One simple way of providing a matched termination is to use a fixed attenuator. For example if a 20 dB attenuator is used then the worst possible case is for the signal transmitted through the attenuator to be totally reflected by either a short circuit or an open circuit. This reflected wave is attenuated by a further 20 dB on passing back through the attenuator. The reflected signal is therefore 40 dB below the incident signal giving a voltage reflection coefficient of 0.01 and a VSWR of 1.02 : 1. This may very well be a better match than the match of the attenuator itself to the incoming wave. An attenuator can be used to improve the match of some other device in just the same way. It is then described as a 'pad'. The disadvantages of using an attenuator as a matched load are that it is unnecessarily bulky and may not be able to dissipate much power.

Matched loads for stripline circuits generally employ one or more lumped-

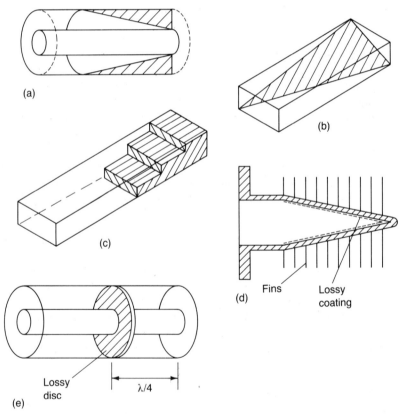

(a)

(b)

(c)

(d)   Fins   Lossy
               coating

(e)   Lossy
      disc   λ/4

Fig. 4.9 Microwave loads: (a) coaxial line, (b) waveguide, (c) compact wavguide, (d) high-power waveguide and (e) compact coaxial line loads.

element pads followed by a lumped resistor whose resistance is equal to the characteristic impedance of the line.

A second approach is based on the one described above. The end of the transmission line or waveguide is filled with a tapered lossy section whose wide end may completely fill the space. Possible materials include iron-loaded epoxy resin and carbon-loaded ceramic. Figures 4.9(a) and (b) show typical forms for use in coaxial line and waveguide. The principle employed is to make the load material as lossy as possible and to rely on the gradual taper to ensure a satisfactory match. For high-precision measurements the load can be left free to slide inside the transmission line or waveguide. As it is moved backwards and forwards the phase of the residual mismatch changes. In this way the mismatch of the load can be distinguished from that of the device under test.

It is possible to make more compact loads by more sophisticated design techniques. One of these is the use of a series of quarter-wave steps in place of the taper as shown in the waveguide load of Fig. 4.9(c). It can be shown that such an arrangement has a better match than a taper of the same length. It also has the advantage that the load itself no longer has a thin, fragile, tip.

High-power loads are sometimes made by tapering the guide itself, coating the interior with a temperature-resistant material such as flame-sprayed Kanthal, and providing cooling fins on the outside. This arrangements is shown in Fig. 4.9(d).

Very compact loads can be made using a sheet of resistive card matched to the wave impedance of the incoming wave. This impedance appears in parallel with the impedance of the transmission line beyond it. By putting a short circuit a quarter wavelength behind the card that impedance is close to an open circuit over a range of frequencies. Figure 4.9(e) shows this arrangement in coaxial line. A similar technique can be used to make compact waveguide loads.

Absorbent materials are manufactured which are designed to absorb TEM waves in free space. They generally take the form of arrays of pyramids which present a gradual transition from the impedance of free space to that of the absorber. These materials are used for lining the anechoic ('echo-free') chambers employed for testing antennas and for electromagnetic compatibility measurements (Keiser, 1979).

4.8 CONCLUSION

In this chapter we have examined the effects on electromagnetic waves of real conducting boundaries which cause some loss. We have seen that metallic boundaries are usually very badly mismatched to the wave impedance of the incoming wave so that there is almost total reflection. The wave impedance of the conductor is complex implying loss in a wave travelling through it and a phase change in the wave reflected from its surface. The combination of reflection loss and transmission loss makes it possible to use thin metal sheets to screen electronic equipment against electromagnetic interference. The reflection of waves which are obliquely incident on a conducting surface was examined leading to a discussion of the attenuation of signals in transmission lines and waveguides. Finally techniques for deliberately introducing loss in the form of attenuators and matched loads were described.

EXERCISES

4.1 Calculate the skin depth and surface resistance for brass ($\sigma = 1.1 \times 10^7\,\mathrm{S\,m^{-1}}$) and gold ($\sigma = 4.1 \times 10^7\,\mathrm{S\,m^{-1}}$) at frequencies of 200 MHz, 2 GHz and 20 GHz.

4.2 It is proposed to make a room secure against electronic 'bugging' at frequency of 100 MHz by coating the windows with a film of copper ($\sigma = 5.7 \times 10^7\,\mathrm{S\,m^{-1}}$) 10 μm thick. If the window glass is 4 mm thick and has a relative permittivity of 4.0 estimate the change in the transmission loss produced by the copper film.

4.3 Estimate the electric and magnetic screening effectiveness of box whose dimensions are 20 mm × 50 mm × 200 mm which is made of brass ($\sigma = 1.1 \times 10^7\,\mathrm{S\,m^{-1}}$) 0.5 mm thick over the frequency range 10 Hz to 100 MHz.

4.4 Estimate the loss per unit length in a section of WG16 waveguide made of copper ($\sigma = 5.7 \times 10^7\,\mathrm{S\,m^{-1}}$) at a frequency of 10 GHz.

Antennas

5.1 INTRODUCTION

In Chapter 1 we saw that electromagnetic waves can propagate through a uniform dielectric medium whose boundaries are so far away that they can be ignored. The subsequent chapters examined the effects of boundaries on wave propagation and, in particular, the properties of waves guided by transmission lines and waveguides. In this chapter we consider how waves in free space can be excited by waves in waveguides and vice versa. Essentially this is a problem of matching between the two media of propagation and a device which performs that function is called an antenna. Antennas are familiar objects in the modern world with its multitude of radio and television aerials (the old word for antennas) and the growing number of dishes for receiving satellite transmissions.

5.2 MAGNETIC VECTOR POTENTIAL

Before proceeding to the theory of antennas we require one new idea: the magnetic vector potential. In elementary textbooks (Carter, 1986) it is shown that the electric field can be calculated from the electrostatic potential using

$$E = -\nabla V \tag{5.1}$$

where

$$\nabla V = \hat{x}\,\frac{\partial V}{\partial x} + \hat{y}\,\frac{\partial V}{\partial y} + \hat{z}\,\frac{\partial V}{\partial z} \tag{5.2}$$

in rectangular Cartesian coordinates. Equation (5.1) can also be interpreted in other coordinate systems as we shall see. Now (5.1) was derived from the theory of electrostatics but it cannot be correct for problems involving electromagnetic waves because it neglects the part of the electric field produced by a changing magnetic field. To get round this problem we define a new vector A, known as the magnetic vector potential, by

$$B = \nabla \wedge A. \tag{5.3}$$

It may be recalled that the magnetic scalar potential defined by an equation analagous to (5.1) is of limited value because it is not a single-valued function (Carter, 1986, p. 56). Substituting (5.3) into (1.8) gives

$$\nabla \wedge E = -\frac{\partial}{\partial t}(\nabla \wedge A) \tag{5.4}$$

so that

$$\nabla \wedge \left(E + \frac{\partial A}{\partial t}\right) = 0. \tag{5.5}$$

Now suppose that the quantity in the brackets is equal to minus the gradient of the electrostatic potential V. Then

$$E = -\nabla V - \frac{\partial A}{\partial t}. \tag{5.6}$$

This equation, therefore, replaces (5.1) and reduces to it when the problem is a static one. The substitution of ∇V is justified because it can be shown that

$$\nabla \wedge (\nabla V) \equiv 0. \tag{5.7}$$

When (5.3) is substituted into (1.7) the result is

$$\nabla \wedge (\nabla \wedge A) = \mu_0 J + \mu_0 \frac{\partial D}{\partial t}, \tag{5.8}$$

where, for simplicity, it has been assumed that only fields in free space are to be considered. It can be shown that the left-hand side of this equation can be rewritten

$$\nabla \wedge (\nabla \wedge A) \equiv \nabla(\nabla \cdot A) - \nabla^2 A. \tag{5.9}$$

In fact A is not completely defined by (5.3) and we are free to impose the additional condition

$$\nabla \cdot A = -\varepsilon_0 \mu_0 \frac{\partial V}{\partial t}. \tag{5.10}$$

This is known as the Lorentz condition. It can be shown to be completely consistent with Maxwell's equations and it has the effect of making A depend solely on the distribution of currents as we shall see.

Substitution into (5.8) from (5.6), (5.9) and (5.10) gives

$$\nabla^2 A - \varepsilon_0 \mu_0 \frac{\partial^2 A}{\partial t^2} = -\mu_0 J. \tag{5.11}$$

This is a wave equation relating the components of A to the corresponding components of the source current.

Similarly, taking the divergence of (5.6) we have

$$\nabla \cdot E = -\nabla^2 V - \frac{\partial}{\partial t}(\nabla \cdot A) = \varrho/\varepsilon_0$$

so that
$$\nabla^2 V - \varepsilon_0\mu_0 \frac{\partial^2 V}{\partial t^2} = -\varrho/\varepsilon_0. \qquad (5.12)$$

Comparison of (5.11) and (5.12) shows that A bears a relationship to current sources very similar to that borne by V to charge sources.

5.3 RETARDED POTENTIALS

Equations (5.11) and (5.12) express the idea that if a current or a charge is changing with time the effect travels outwards with the speed of light. Thus if we are examining the electrostatic potential at a point P distant r from a charge Q the potential at time t will depend upon the value of Q at the earlier time $(t - r/c)$. This value of the charge is referred to as the retarded value and is denoted by $[Q]$. Using this notation the solution of (5.12) can be written

$$V = \frac{1}{4\pi\varepsilon_0} \frac{[Q]}{r} \qquad (5.13)$$

for a point charge. For a distribution of charge this equation must be integrated over the space containing the charge to give

$$V = \frac{1}{4\pi\varepsilon_0} \iiint \frac{[\varrho]}{r} \, dv, \qquad (5.14)$$

where the distance r ranges over the volume v as the integration is carried out.

By analogy the magnetic vector potential is given by

$$A = \frac{\mu_0}{4\pi} \iiint \frac{[J]}{r} \, dv. \qquad (5.15)$$

These ideas are put to use in the next section.

5.4 SMALL ELECTRIC DIPOLE

Figure 5.1 shows one possible way of launching waves from the end of a coaxial line. The outer conductor of the line is connected to the edge of a hole in a large conducting plane whilst the end of the inner conductor projects a short distance into the space beyond the plane, where it is

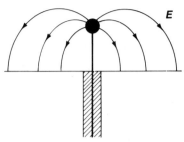

Fig. 5.1 The electric field of a monopole antenna formed by allowing the central conductor of a coaxial line to project through a conducting plane.

terminated by a small sphere. In this context 'short' means 'short compared with the free-space wavelength'. The end of the line is a capacitor whose electric field pattern is somewhat as shown in Fig. 5.1. If is assumed that all the charges accumulate on the spherical end then the current in the connecting wire can be considered to be uniform along its length and varying sinusoidally with time.

To analyse the antenna shown in Fig. 5.1 we employ the method of images and replace the lower half of the diagram by an identical antenna with opposite polarity as shown in Fig. 5.2. This arrangement with its pair of positive and negative charges is known as a dipole. The figure shows the electric and magnetic fields around the dipole at a moment when the polarities of the charges and the current in the wire are as shown. Examination of the directions of E and H shows that the Poynting vector is directed outwards. In the next quarter cycle, however, the direction of H is reversed and the Poynting vector points inwards suggesting that the field around the dipole stores energy but does not radiate it. We therefore have to show that the dipole does indeed radiate.

The field pattern around the dipole is cylindrically symmetrical. At large distances from it the length of the dipole is insignificant and the problem has spherical symmetry. It is best, therefore, to examine the problem in the spherical polar co-ordinate system shown in Fig. 5.3 (see Appendix B).

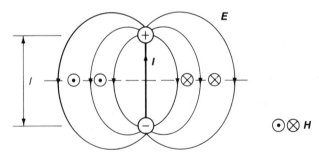

Fig. 5.2 The electric field around an alternating electric dipole.

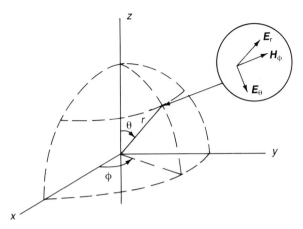

Fig. 5.3 Spherical polar coordinates.

From the symmetry of the problem we expect that the electric field will
have r and θ components and the magnetic field only a ϕ component as
shown in the inset to Fig. 5.3.

The dipole is represented by a pair of oscillating charges

$$q = \pm q_0 e^{j\omega t} \tag{5.16}$$

located at $\pm l/2$ as shown in Fig. 5.4. The potential at P is then

$$V = \frac{1}{4\pi\varepsilon_0} \left[\frac{q_0 e^{j(\omega t - k_0 r_1)}}{r_1} - \frac{q_0 e^{j(\omega t - k_0 r_2)}}{r_2} \right], \tag{5.17}$$

where r_1 and r_2 are the distances from the two charges to P. If $r \gg l$ we can
write

$$r_1 = r - \tfrac{1}{2}l \cos \theta = r - \delta r$$

and
$$r_2 = r + \tfrac{1}{2}l \cos \theta = r + \delta r \tag{5.18}$$

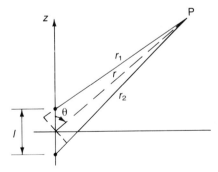

Fig. 5.4 Geometry of the radiation from a small electric dipole antenna.

so that

$$V = \frac{q_0}{4\pi\varepsilon_0 r} \left[\frac{e^{jk_0\delta r}}{(1 - \delta r/r)} - \frac{e^{-jk_0\delta r}}{(1 + \delta r/r)} \right] e^{j(\omega t - k_0 r)}. \qquad (5.19)$$

Expanding the terms as power series and neglecting terms above the first order in $(\delta r/r)$ we obtain

$$V \simeq \frac{q_0}{4\pi\varepsilon_0 r} \left[\left(1 + \frac{\delta r}{r}\right)\left(1 + jk_0\delta r\right) - \left(1 - \frac{\delta r}{r}\right)\left(1 - jk_0\delta r\right) \right] e^{j(\omega t - k_0 r)}$$

$$\simeq \frac{q_0 l \cos\theta}{4\pi\varepsilon_0} \left(\frac{1}{r^2} + j\frac{k_0}{r}\right) e^{j(\omega t - k_0 r)} \qquad (5.20)$$

which can be written

$$V \simeq \frac{[p] \cos\theta}{4\pi\varepsilon_0} \left(\frac{1}{r^2} + j\frac{k_0}{r}\right), \qquad (5.21)$$

where $p = q_0 l$ is the electric dipole moment. The first term in (5.21) falls off more rapidly with r than does the second one. It represents the field of an electrostatic dipole. The second term arises because of the difference between the propagation times from the two charges to the point P.

The magnetic vector potential of the dipole is from (5.15)

$$A = \frac{j\omega\mu_0[p]}{4\pi r} \qquad (5.22)$$

because

$$\iiint J \, dv = Il = \frac{dq}{dt} l = j\omega p. \qquad (5.23)$$

The electric field of the dipole is obtained by substituting the electric and magnetic potentials from (5.21) and (5.22) into (5.6). It is convenient to consider separately the contributions from the two potentials. In spherical polar coordinates

$$\nabla V = \hat{r} \frac{\partial V}{\partial r} + \frac{\hat{\theta}}{r} \frac{\partial V}{\partial \theta} + \frac{\hat{\phi}}{r \sin\theta} \frac{\partial V}{\partial \phi}, \qquad (5.24)$$

(see Appendix B) where $\hat{r}$, $\hat{\theta}$ and $\hat{\phi}$ are unit vectors in the three co-ordinate directions. Thus the electric potential contributes the field components

$$E_r = \frac{[p] \cos\theta}{4\pi\varepsilon_0} \left(\frac{2}{r^3} + \frac{2jk_0}{r^2} - \frac{k_0^2}{r}\right)$$

$$E_\theta = \frac{[p] \sin\theta}{4\pi\varepsilon_0} \left(\frac{1}{r^3} + j\frac{k_0}{r^2}\right)$$

$$E_\phi = 0. \qquad (5.25)$$

The vector A is in the z direction so it has components

$$A_r = A \cos \theta$$
$$A_\theta = -A \sin \theta$$
$$A_\phi = 0, \tag{5.26}$$

giving contributions to the electric field from the magnetic potential

$$E_r = \frac{k_0^2[p]}{4\pi\varepsilon_0 r} \cos \theta$$
$$E_\theta = -\frac{k_0^2[p]}{4\pi\varepsilon_0 r} \sin \theta \tag{5.27}$$

so that, finally

$$E_r = \frac{[p] \cos \theta}{2\pi\varepsilon_0} \left(\frac{1}{r^3} + j\frac{k_0}{r^2} \right)$$
$$E_\theta = \frac{[p] \sin \theta}{4\pi\varepsilon_0} \left(\frac{1}{r^3} + j\frac{k_0}{r^2} - \frac{k_0^2}{r} \right). \tag{5.28}$$

The magnetic field is obtained from (5.3) and (5.22) making use of the expression for the curl of A in spherical polar co-ordinates (see Appendix B)

$$\nabla \wedge A = \frac{\hat{r}}{r \sin \theta} \left[\frac{\partial}{\partial \theta} (A_\phi \sin \theta) - \frac{\partial A_\theta}{\partial \phi} \right]$$
$$+ \frac{\hat{\theta}}{r} \left[\frac{1}{\sin \theta} \frac{\partial A_r}{\partial \phi} - \frac{\partial}{\partial r} (rA_\phi) \right]$$
$$+ \frac{\hat{\phi}}{r} \left[\frac{\partial}{\partial r} (rA_\theta) - \frac{\partial A_r}{\partial \theta} \right]. \tag{5.29}$$

Fortunately most of the terms in this fearsome expression are zero because A has only r and θ components and these only vary with θ. Thus the only component of H is

$$H_\phi = \frac{1}{\mu_0 r} \left[\frac{\partial}{\partial r}(rA_\theta) - \frac{\partial A_r}{\partial \theta} \right]$$
$$= j\frac{\omega[p] \sin \theta}{4\pi} \left[\frac{1}{r^2} + j\frac{k_0}{r} \right]. \tag{5.30}$$

We recall that $j\omega p = Il$.

The complete expressions for the field of the dipole are rather unwieldy so it is useful to consider them in three parts. We note that some of the terms fall off with increasing r much faster than others. The terms which fall off fastest are

$$E_r = \frac{[p] \cos \theta}{2\pi\varepsilon_0 r^3}$$

$$E_\theta = \frac{[p] \sin \theta}{4\pi\varepsilon_0 r^3}. \qquad (5.31)$$

These correspond to the electrostatic field of the dipole and represent a capacitive storage of energy.

The next terms are

$$E_r = j\frac{k_0[p] \cos \theta}{2\pi\varepsilon_0 r^2}$$

$$E_\theta = j\frac{k_0[p] \sin \theta}{4\pi\varepsilon_0 r^2}$$

$$H_\phi = j\,\frac{\omega[p] \sin \theta}{4\pi r^2}. \qquad (5.32)$$

The magnetic field is just that given by the Biot–Savart law for a current element (Carter, 1986, p. 53), so this term also represents a quasi-static field. If the direction of the Poynting vector for the fields in (5.32) is examined it is found that the energy associated with them is circulating close to the dipole. These terms are known as the induction field. Together with the electrostatic field they form the near field of the dipole.

The remaining terms are the far field

$$E_\theta = \frac{-k_0^2[p] \sin \theta}{4\pi\varepsilon_0 r}$$

$$H_\phi = \frac{-\omega k_0[p] \sin \theta}{4\pi r}. \qquad (5.33)$$

These are in phase with each other and the Poynting vector is directed radially outwards indicating that they represent radiation of power by the dipole. The wave impedance is

$$Z_0 = \frac{E_\theta}{H_\phi} = \frac{k_0}{\omega\varepsilon_0} = \sqrt{\left(\frac{\mu_0}{\varepsilon_0}\right)} \qquad (5.34)$$

exactly as for a plane TEM wave. In fact such a plane wave can be thought of as the limiting case of a spherical wave at large distances from the antenna.

The time average value of the Poynting vector is

$$S_r = \frac{1}{r}\left(\frac{p}{4\pi r}\right)^2 Z_0\omega^2 k_0^2 \sin^2 \theta. \qquad (5.35)$$

To find the radiated power we integrate this over the surface of a sphere. The annular element of area shown in Fig. 5.5 has radius $r \sin \theta$ and width $r\,d\theta$. Its area is therefore $2\pi r^2 \sin \theta\,d\theta$. The power flow is given by

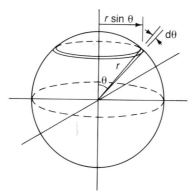

Fig. 5.5 Geometry for integrating the Poynting vector of a small dipole radiator over a sphere.

$$P = \int_0^\pi S_r \, 2\pi r^2 \sin \theta \, d\theta$$

$$= \frac{\omega^2 k_0^2 p^2 Z_0}{12\pi} = \frac{k_0^2 I^2 l^2 Z_0}{12\pi}. \tag{5.36}$$

This expression is independent of the radius of the sphere as we would expect since the total power flow is the same through any closed surface surrounding the dipole. Equation (5.36) can be expressed in a slightly different form by recalling that $k_0 = 2\pi/\lambda$

$$P = 40\pi^2 I^2 (l/\lambda)^2. \tag{5.37}$$

The input resistance of the dipole is, therefore,

$$R_r = \frac{2P}{I^2} = 80\pi^2 (l/\lambda)^2. \tag{5.38}$$

This is known as the radiation resistance of the dipole, it can be seen that it increases with length. If we assume that $(l/\lambda) = 0.1$ which is about the maximum value for which the theory is valid then $R_r = 7.9\,\Omega$. This figure presents two problems. First it is not well matched to typical transmission-line impedances and, second, it means that a transmitter must supply large currents at low voltage giving rise to large ohmic losses in the connecting cables. The near field of the antenna stores electromagnetic energy and, therefore, contributes a reactive component to the input impedance (see Jordan and Balmain, 1968).

If the power radiated from the dipole were distributed uniformly over the surface of a sphere of radius r then the magnitude of the Poynting vector would be

$$S_i = \frac{1}{3} \left(\frac{p}{4\pi r} \right)^2 Z_0 \omega^2 k_0^2. \tag{5.39}$$

This is the power density radiated by an isotropic ('same in all directions') antenna. The directivity of an antenna is defined by comparing the magnitude of the Poynting vector in a given direction with the magnitude of the Poynting vector of an isotropic antenna (with the same input power) at the same position in space. Thus the directivity is given by

$$D(\theta, \phi) = \frac{S_r(r, \theta, \phi)}{S_i(r)}. \tag{5.40}$$

The directivity of a small dipole is 1.5 from (5.35) and (5.39). This may also be expressed in decibels as 1.76 dBi where the symbol dBi indicates that the reference signal is that of an isotropic radiator. Alternatively directivity may be defined relative to the radiation pattern of a standard antenna with the same input power.

Not all the power input to an antenna is radiated. Some of it is dissipated in the antenna. The ratio of the radiated power to the input power is the radiation efficiency of the antenna

$$\eta = \frac{P_r}{P_{in}}. \tag{5.41}$$

The gain of an antenna relates its radiation pattern to the input power. Thus

$$G(\theta, \phi) = \eta D(\theta, \phi). \tag{5.42}$$

The difference between the gain and the directivity of an antenna is that the former takes account of losses within the antenna whilst the latter does not. The word 'gain' is sometimes used, loosely, in place of 'directivity'. Although the gain and directivity of an antenna can be specified in any

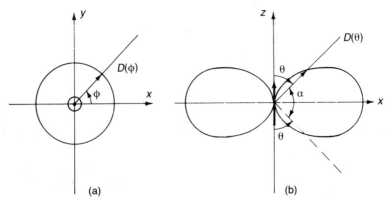

Fig. 5.6 Polar radiation diagrams for a small electric dipole antenna: (a) in a plane perpendicular to the dipole and (b) in a plane containing the dipole.

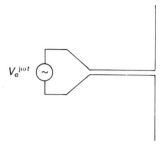

Fig. 5.7 Arrangement of a centre-fed dipole antenna.

direction the usual practice is to quote the figures corresponding to the direction in which the power density is maximum.

The directional properties of antennas are usually displayed as polar plots of the directivity. Figure 5.6 shows the polar plots for a small dipole in planes perpendicular to and parallel to the dipole. A useful measure of the directional properties of an antenna is the angle at which the power density is half the maximum value, that is

$$S_r(\theta) = \tfrac{1}{2}(S_r)_{max}. \tag{5.43}$$

For a small dipole $S_r(\theta) \propto \sin^2 \theta$ so that, from (5.43), $\theta = 45°$. The included angle (α in Fig. 5.5(b)) is known as the half-power beamwidth. For a small dipole it is 90°.

So far we have assumed that the dipole is fed by a coaxial cable as shown in Fig. 5.1. A very common alternative is the centre-fed dipole shown in Fig. 5.7 in which the feeder is a two-wire line.

5.5 THE RECIPROCITY THEOREM

So far we have considered antennas as radiating elements. They can, equally well, act as receivers of electromagnetic radiation. The properties of an antenna as a transmitter and a receiver are linked by the reciprocity theorem which states that

In a linear system the response at a point to a stimulus at another point is unchanged when the stimulus and response are exchanged.
(Carter and Richardson, 1972)

An example of a linear system is that comprising a pair of antennas as shown in Fig. 5.8(a). Provided that the antennas and the medium surrounding them are linear, passive and isotropic then the system shown in the figure can be represented by the matrix equation

$$\begin{pmatrix} I_1 \\ I_2 \end{pmatrix} = \begin{pmatrix} Y_{11} & Y_{12} \\ Y_{21} & Y_{22} \end{pmatrix} \begin{pmatrix} V_1 \\ V_2 \end{pmatrix}. \tag{5.44}$$

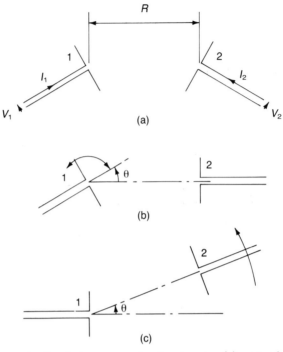

Fig. 5.8 Communication between a pair of antennas: (a) general arrangement, (b) when antenna 1 is rotated and (c) when antenna 2 is rotated.

The reciprocity theorem implies that the relationship between V_2 and I_1 is the same as that between V_1 and I_2 so that $Y_{12} = Y_{21}$. Thus an alternative statement of the reciprocity theorem is

> The impedance and admittance matrices of passive linear systems are symmetrical.

As applied to the pair of antennas shown in Fig. 5.8(a) the effect of the theorem is that the behaviour of the system is unchanged when the transmitting and receiving antennas are exchanged. Now the directivity of antenna 1 can be measured either by keeping antenna 2 fixed and rotating antenna 1 or by keeping antenna 1 fixed and moving antenna 2 as shown in Figs. 5.8(b) and (c). Thus the directivity of an antenna is the same whether it is transmitting or receiving.

When an antenna is acting as a receiver a useful measure of its effectiveness is the area from which it gathers power. The effective aperture is defined by

$$A_e = \frac{\text{Power absorbed by load}}{\text{Power density in incident wave}}. \tag{5.45}$$

To find the relationship between the effective aperture and the gain of an antenna we consider a pair of antennas as shown in Fig. 5.8(a). We shall assume that the separation between the antennas is large enough for the wave at the receiving antenna to be effectively that of a uniform plane wave. The power density is then

$$S = g_1 \frac{P_{in}}{4\pi R^2}, \tag{5.46}$$

where P_{in} is the input power and g_1 the gain of antenna 1. Then, from (5.45) the power absorbed by the load is

$$P_L = A_{e2}S = \frac{A_{e2}g_1}{4\pi R^2} P_{in}. \tag{5.47}$$

Now the reciprocity theorem tells us that the relationship between the input power and the load power must be the same if the roles of the antennas are exchanged so

$$P_L = \frac{A_{e1}g_2}{4\pi R^2} P_{in} \tag{5.48}$$

and therefore

$$\frac{A_{e1}}{A_{e2}} = \frac{g_1}{g_2} \tag{5.49}$$

so that the effective aperture of an antenna is proportional to the gain. The constant of proportionality may be found by considering a small lossless dipole acting as a receiving antenna as shown in Fig. 5.9(a). The voltage induced in the antenna is $E_i l$. The equivalent circuit of the dipole connected to the load is shown in Fig. 5.9(b). The maximum power theorem requires $Z_L = R_r$ for maximum power transfer to the load. The load power is then

$$P_L = \frac{1}{2} \left(\frac{V}{2}\right)^2 \frac{1}{R_r} = \frac{E_i^2 l^2}{8R_r}. \tag{5.50}$$

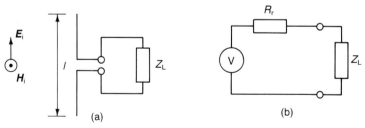

(a)   (b)

Fig. 5.9 Dipole receiving antenna: (a) general arrangement and (b) equivalent circuit.

But the power density in the incident wave is

$$S = \frac{E_i^2}{2Z_0} \qquad (5.51)$$

so, from (5.45)

$$A_e = \frac{Z_0}{4R_r} l^2. \qquad (5.52)$$

Substituting for the radiation resistance of the dipole from (5.38) we have

$$A_e = \frac{\lambda^2}{4\pi} \frac{3}{2}. \qquad (5.53)$$

Therefore, since the gain of a small dipole is 1.5, we conclude that, for any antenna

$$A_e = \frac{\lambda^2}{4\pi} g. \qquad (5.54)$$

5.6 SMALL MAGNETIC DIPOLE

The symmetry of Maxwell's equations in free space means that, if the fields $E(r, t)$ and $H(r, t)$ are a solution of them, then the fields

$$E'(r, t) = -\frac{1}{\varepsilon_0} H(r, t)$$

and $\qquad H'(r, t) = \frac{1}{\mu_0} E(r, t) \qquad (5.55)$

are also a solution. This is easily demonstrated by substituting (5.55) into (1.7) and (1.8). The fields given by (5.55) are called the duals of the original fields and their existence demonstrates the principle of duality. This principle allows us to deduce the solution to one problem from that for its dual.

The dual of the electric dipole considered in Section 5.4 is the magnetic dipole. The fields around it can be derived by applying principle of duality to the fields of the electric dipole with the results

$$H_r = \frac{[p]}{2\pi\varepsilon_0\mu_0} \left(\frac{1}{r^3} + j\frac{k_0}{r^2}\right) \cos\theta \qquad (5.56)$$

$$H_\theta = \frac{[p]}{4\pi\varepsilon_0\mu_0} \left(\frac{1}{r^3} + j\frac{k_0}{r^2} - \frac{k_0^2}{r}\right) \sin\theta \qquad (5.57)$$

$$E_\phi = \frac{-[p]}{4\pi\varepsilon_0} \left(\frac{j\omega}{r^2} - \frac{\omega k_0}{r}\right) \sin\theta. \qquad (5.58)$$

The near field of this dipole is

$$H_r = \frac{[p] \cos \theta}{2\pi\varepsilon_0\mu_0 r^3} \qquad (5.59)$$

$$H_\theta = \frac{[p] \sin \theta}{4\pi\varepsilon_0\mu_0 r^3} \qquad (5.60)$$

which is just the field of a magnetic dipole whose dipole moment is

$$[j] = \frac{[p]}{\varepsilon_0\mu_0} \qquad (5.61)$$

as is shown in texts on magnetostatics (Bleaney and Bleaney, 1976). A magnetic dipole is realised as a small current loop as shown in Fig. 5.10 with

$$j = \pi a^2 I. \qquad (5.62)$$

The loop must be small enough for there to be no appreciable phase difference between the currents in different parts of the loop. The far field of a magnetic dipole antenna is therefore

$$H_\theta = -\frac{[j] k_0^2 \sin \theta}{4\pi r} \qquad (5.63)$$

$$E_\phi = \frac{\mu_0 [j] \omega k_0 \sin \theta}{4\pi r}. \qquad (5.64)$$

The reversal in the sign of one of these equations compared with (5.33) ensures that the direction of the Poynting vector is outwards. By analogy with (5.36) the total radiated power is

$$P = \frac{1}{12\pi} (Z_0 k_0^4 j^2), \qquad (5.65)$$

where j is given by (5.62). If the dipole has the form of the current loop shown in Fig. 5.8 then the radiation resistance is

$$R_r = \frac{8\pi Z_0}{3} \left(\frac{\pi a}{\lambda}\right)^4. \qquad (5.66)$$

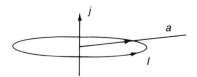

Fig. 5.10 Small magnetic dipole antenna.

Comparing this expression with (5.38) we see that the radiation resistance of a magnetic dipole varies much more rapidly with frequency than that of an electric dipole. If the diameter of the loop is 0.1λ then the radiation resistance is $1.92\ \Omega$. This is an inconveniently small impedance but it can easily be increased by using a coil of several turns. For N turns the radiation resistance increases as N^2. A further increase can be obtained by winding the coil on a ferrite rod. This is the kind of antenna usually used in portable radios.

5.7 HALF-WAVE DIPOLE

The inconveniently small radiation resistance of an electric dipole can be increased by making it longer so that it resonates. If each arm of the dipole is a quarter wavelength long then there is a standing wave and the current varies sinusoidally along it as shown in Fig. 5.11. When this antenna is compared with that shown in Fig. 5.2 it is seen that the charges accumulate along the length of the dipole rather than being concentrated at its ends. This antenna can be analysed by assuming that it is made up of a large number of small dipoles as shown in Fig. 5.11(a). At large distances from the antenna the path distance is effectively the same for all elements as far as the signal amplitude is concerned. For the element shown the phase is

$$\exp j[\omega t - k_0(R - x \sin \theta)] \qquad (5.67)$$

and the dipole moment is

$$p = -j\frac{I\,dx}{\omega} \cos k_0 x. \qquad (5.68)$$

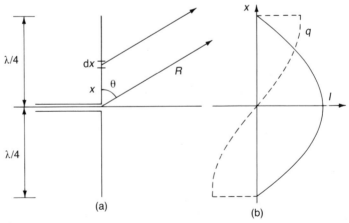

Fig. 5.11 Half-wave dipole antenna: (a) geometry and (b) current and charge distributions assumed.

The variation cannot be exactly sinusoidal because of the radiation of energy from the antenna but careful experiments have shown that the departure from the distribution assumed is insignificant. The electric field at large distances from the antenna is

$$E_\theta = -j\frac{k_0^2 \sin\theta\, I}{4\pi\varepsilon_0\omega R} \int_{-\lambda/4}^{\lambda/4} \cos k_0 x\; e^{j[\omega t - k(R - x\sin\theta)]}\, dx. \qquad (5.69)$$

This expression can be integrated to give

$$E_\theta = \frac{jZ_0}{2\pi R} \frac{\cos\left[(\pi/2)\cos\theta\right]}{\sin\theta} [I]. \qquad (5.70)$$

This field pattern is very like that of the short dipole. It has cylindrical symmetry, maximum field in the plane normal to the antenna and zero radiation along the axis of the antenna. The half-power beam width is 78° and the maximum directivity is 1.64, figures which do not differ markedly from those for a short dipole. The big difference is in the radiation resistance which is 73 Ω for a half-wave dipole. This is much more practical and half-wave dipoles are in common use for television receiving antennas. Incidentally this figure explains why the characteristic impedance of the cable used for television aerial downleads is 75 Ω rather than the usual 50 Ω.

An arrangement which is sometimes used is a monopole antenna combined with a ground plane along the lines of Fig. 5.1. This has a radiation resistance which is just half of that of the corresponding dipole and has double the directivity because the radiation is concentrated over one hemisphere. Short monopoles are used for car radio aerials using the bodywork of the car as the ground plane. Quarter-wave monopoles are used for VHF radio transmitters with the conductivity of the earth supplying the ground plane.

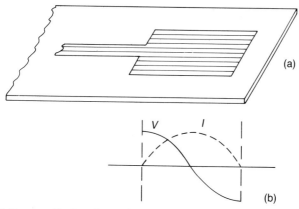

Fig. 5.12 Half-wave dipole microstrip antenna.

Dipole antennas can be realized in microstrip. Figure 5.12 shows one possible arrangement. A stripline of normal characteristic impedance feeds a much wider section whose impedance is therefore much lower. Because of the big difference in the impedances the feed line looks very much like an open circuit to the wider line which then acts as a resonator (see Chapter 7 for a discussion of resonators). The standing wave excited has sinusoidal voltage and current distributions as shown and the section of line therefore acts as a dipole antenna. Further information about microstrip antennas is given by Rudge *et al.* (1982–3).

5.8 DIPOLE ARRAYS

The single dipole antennas discussed so far have the disadvantage that they are not strongly directional. For many purposes it is desirable to radiate the greater part of the power in a narrow beam or to have a receiving antenna with a large effective aperture. This increase in directivity can be achieved by using more than one dipole. The number of possible arrangements is very large so here we shall concentrate on illustrating the basic principles involved.

Figure 5.13 shows two half-wave dipoles arranged parallel to each other and d apart. Each dipole radiates uniformly in all directions in the plane of the paper. We will assume that the dipoles are driven with currents of equal amplitude and relative phase α. Then the phase difference between the signals from the two dipoles at a distant point in the plane is

$$\psi = k_0 d \cos \phi + \alpha. \tag{5.71}$$

The total field intensity at that point is

$$\begin{aligned} E &= E_0 \exp j\psi/2 + E_0 \exp -j\psi/2 \\ &= 2E_0 \cos \psi/2, \end{aligned} \tag{5.72}$$

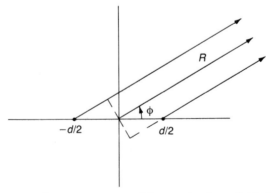

Fig. 5.13 Geometry of an antenna comprising a pair of parallel half-wave dipoles.

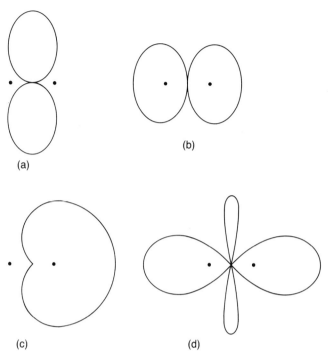

Fig. 5.14 Radiation patterns of antennas comprising pairs of half-wave dipoles having the following spacings and phases: (a) $\lambda/2$, zero, (b) $\lambda/2$, 180°, (c) $\lambda/4$, −90° and (d) λ, zero (after Jordan and Balmain, 1968).

where E_0 is the field strength due to one dipole on its own. Different radiation patterns can be produced by various choices of the separation between the dipoles and of the phase difference between them. Figure 5.14 shows a number of examples. These show how the direction of the maximum directivity can be steered by changing the relative phase of the signals fed to the dipoles. Figure 5.14(c) is interesting because it shows a radiation pattern which does not have a backward lobe. Figure 5.14(d) shows the presence of sidelobes in addition to the main lobes.

The cardioid radiation pattern of Fig. 5.14(c) is given by

$$E = 2E_0 \cos[(\cos \phi - 1)\pi/4]. \tag{5.73}$$

This has a maximum value of 2 in the $\phi = 0$ direction because the signals from the two dipoles are in phase. The maximum directivity of the antenna is thus four times that of one dipole that is 6.56 if half wave dipoles are used.

By adding more dipoles we can get greater directivity. The field at a remote point due to the array of N dipoles shown in Fig. 5.15 is

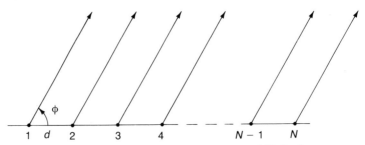

Fig. 5.15 Geometry for radiation from a linear array of N dipoles.

$$E = E_0(1 + e^{j\psi} + e^{2j\psi} + \cdots + e^{(N-1)j\psi}), \tag{5.74}$$

where ψ is given by (5.71) as before. The series in (5.74) is a geometrical progression whose sum is

$$E = E_0 \left(\frac{1 - e^{Nj\psi}}{1 - e^{j\psi}} \right) \tag{5.75}$$

so

$$|E| = |E_0| \left| \frac{\sin N\psi/2}{\sin \psi/2} \right| \tag{5.76}$$

which has a maximum value of $N|E_0|$ when $\psi = 0$. The actual direction in space of the maximum radiation depends upon the relative phases of the dipoles. The direction of maximum directivity is given by

$$\psi = k_0 d \cos \phi + \alpha = 0. \tag{5.77}$$

If $\alpha = -k_0 d$ then $\phi = 0$ and the maximum directivity is in the direction of the array and the antenna is described as an endfire array. If the dipoles are fed in phase with each other $\phi = 90°$ and the maximum signal is in a direction at right angles to the array which is then called a broadside array.

The function in (5.76) has zeroes when

$$\psi = \frac{2\pi}{N}, \frac{4\pi}{N}, \frac{6\pi}{N}, \text{ etc.} \tag{5.78}$$

and sidelobe maxima when

$$\psi = \frac{3\pi}{N}, \frac{5\pi}{N}, \frac{7\pi}{N}, \text{ etc.} \tag{5.79}$$

Figure 5.16 shows the polar diagram for the electric field strength of a six-element array. It must be remembered that this diagram only shows the radiation pattern in the plane normal to the dipoles. In planes at right angles to this one the pattern is that of an individual dipole scaled by the

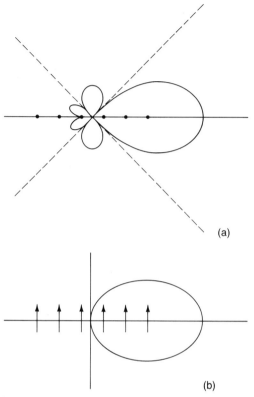

(a)

(b)

Fig. 5.16 Polar diagrams for radiation from a six element linear array: (a) in the plane bisecting the dipoles at right angles and (b) in the plane containing the dipoles.

variation given by (5.76). This is rather difficult to represent graphically so it is usual to display the radiation patterns in principal planes. Figure 5.16(b) shows the radiation pattern in the direction of the main lobe and in a plane parallel to the dipoles.

A very common example of an array is the Yagi–Uda array (Rudge *et al.*, 1982–3), shown in Fig. 5.17, which is used for television reception. The array has only one active element, normally a half-wave dipole. All the other dipoles are parasitic elements excited by the electromagnetic field. The spacing between the elements is only a small fraction of a wavelength. Since a conductor cannot have an electric field tangential to its surface each parasitic dipole must be excited in antiphase with the exciting field. The power extracted from the incident wave is then re-radiated and the complete field of the antenna is found by superimposing the fields of the dipoles. A closely spaced pair of dipoles excited in antiphase has an endfire field pattern so, by extension, a Yagi–Uda array is an endfire array. If such an

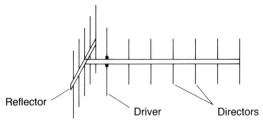

Fig. 5.17 Arrangement of a Yagi–Uda antenna.

array were arranged with a conducting plane normal to its axis and close to the active dipole the result would be an increase in the directivity because of the reflection from the plane. A practical antenna uses either a small array of parasitic dipoles (as shown in Fig. 5.17) or sometimes just one slightly longer dipole to achieve the same effect.

5.9 RADIATION FROM APERTURES

A radiating dipole can be thought of as a termination of a two-wire line which matches the wave on the line to the radiation field. In a similar way we can imagine antennas which are based on matching the fields in a hollow waveguide to the radiation field.

The basis of the analysis of radiation from apertures is Huygens' theory of secondary sources. According to this theory a propagating wavefront can be regarded as being made up of a large number of secondary sources each radiating a spherical wave as shown in Fig. 5.18. The superposition of these waves forms a new wavefront and so on. It turns out that, in order to model this process correctly each secondary source should have a radiation pattern which varies as $(1 + \cos \theta)$, where θ is the angle measured from the normal to the wavefront, and a phase which is 90° ahead of that of the wavefront (Jordan and Balmain, 1968). Note that the angular variation ensures that there is no radiation back towards the source.

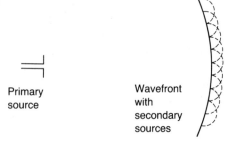

Fig. 5.18 Representation of wave propagation by superposition of secondary wavelets.

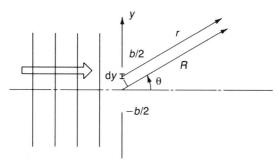

Fig. 5.19 Geometry for radiation from a uniformly illuminated slot.

The simplest example of radiation from an aperture is that which occurs when a gap in a conducting plane is illuminated by a plane wave as shown in Fig. 5.19. The field at a remote point due to the secondary source dy is given by

$$dE = \frac{jk_0E_0a\,dy}{4\pi}(1 + \cos\theta)\frac{e^{-jk_0r}}{r} \qquad (5.80)$$

where a is the width of the aperture in the x direction and

$$r = R - y\sin\theta. \qquad (5.81)$$

The field radiated from the aperture is thus

$$E = \frac{jk_0E_0a(1 + \cos\theta)}{4\pi}\int_{-b/2}^{b/2}\frac{e^{-jk_0(R - y\sin\theta)}}{R - y\sin\theta}\,dy. \qquad (5.82)$$

If $R \gg b$ the bottom line of the integral is approximately equal to R and

$$|E| = \frac{k_0E_0a(1 + \cos\theta)}{4\pi R}\left|\int_{-b/2}^{b/2}\exp(jk_0y\sin\theta)dy\right| \qquad (5.83)$$

which can be integrated to give

$$|E| = \frac{k_0E_0ab}{4\pi R}(1 + \cos\theta)\left[\frac{\sin(\tfrac{1}{2}k_0b\sin\theta)}{\tfrac{1}{2}k_0b\sin\theta}\right]. \qquad (5.84)$$

The angular variation of this expression is dominated by the last term. The dependence of the power density on θ is shown in Fig. 5.20, where $u = (k_0b/2)\sin\theta$. The field nulls are given by

$$\tfrac{1}{2}k_0b\sin\theta = \pm N\pi, \qquad (5.85)$$

where $N = 1, 2, 3$, etc. Hence

$$\theta = \pm\sin^{-1}\left(\frac{N\lambda}{b}\right). \qquad (5.86)$$

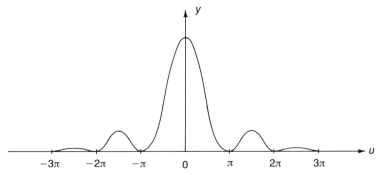

Fig. 5.20 Variation of intensity of radiation from a uniformly illuminated slot with the parameter $u = (k_0 b/2) \sin \theta$.

Clearly the maximum number of nulls is given by

$$N = b/\lambda \qquad (5.87)$$

so that a narrow aperture will produce only a single lobe in the radiation pattern. Conversely a wide aperture produces many lobes with a narrow main lobe. It is generally true that the beam width in a particular plane is inversely proportional to the width of the aperture in that plane. Thus the parabolic reflector shown in Fig. 5.21 has a beam which is narrow in the horizontal direction and wide in the vertical direction somewhat as shown.

In general for an aperture illuminated by a wave whose intensity varies as $E = E_0(y)$ we have

$$E = \frac{jk_0 a(1 + \cos \theta)}{4\pi R} \, e^{-jk_0 R} \int_{-b/2}^{b/2} E_0(y) \, e^{+jk_0 y \sin \theta} \, dy. \qquad (5.88)$$

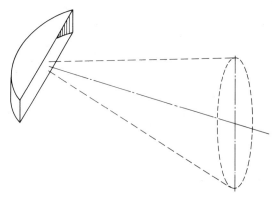

Fig. 5.21 Illustrating the relationship between the dimensions of a parabolic horn antenna and the horizontal and vertical beamwidth.

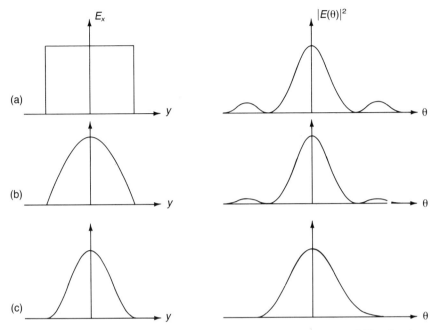

Fig. 5.22 Radiation patterns for a slot with various distributions of illumination: (a) uniform, (b) cosine and (c) cosine squared illumination.

The integral is the Fourier transform of the source distribution. Figure 5.22 shows some source distributions and the corresponding radiation patterns. The first two are those for a uniformly illuminated aperture and for radiation from the open end of a rectangular waveguide. Now the width of a standard waveguide is of the order of magnitude of the free-space wavelength so the width of the beam radiated from an open end is large. To obtain a narrower beam we expand the end of the waveguide into a 'horn' as shown in Fig. 5.23. Provided that the expansion is gradual and the aspect ratio of the horn is the same as that of the waveguide there will be a negligible mismatch between them. Moreover the wave impedance is given by

$$Z_{\mathrm{w}} = Z_0 \frac{\lambda_{\mathrm{g}}}{\lambda_0} \tag{5.89}$$

from (2.45) and (2.46). As the wave travels down the horn λ_{g} tends to λ_0 and the wave impedance tends to that of free space. Thus the horn serves as a means of matching the fields of the waveguide to those of the radiated wave. A more careful examination shows that the wavefront in the horn is cylindrical so that there are small mismatches at the transitions from a plane wave and to a spherical wave. It is also necessary to remember that the surface at the mouth of the horn on which the phase is constant is a

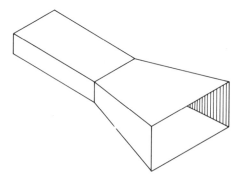

Fig. 5.23 A waveguide horn antenna.

cylinder and not a plane. The $\cos^2$ illumination shown in Fig. 5.22(c) is commonly used in radar antennas.

Another common kind of aperture antenna is the parabolic reflector illuminated by a horn. This antenna and variations on it are used extensively for point-to-point and satellite communication links. Small dishes are becoming increasingly common with the growth of direct broadcasting by satellite. If the horn generates a spherical wave whose apparent source is at the focus of the parabola then the waves reflected from the dish have plane wavefronts. The radiation pattern of the dish may therefore be derived by assuming illumination of its aperture by plane waves. The theory is slightly more involved because the aperture is circular rather than rectangular but the general principles are the same. The directivity of a circular parabolic dish is given by

$$D_{\max} = K\left(\frac{2\pi a}{\lambda}\right)^2 \tag{5.90}$$

where K is a constant in the range $0.61 \leqslant K \leqslant 0.865$. Its effective area is typically around half of its physical area.

5.10 SLOT ANTENNAS

In Chapter 3 we saw that one possible TEM transmission line is the slot line formed by the gap between two parallel plates. If a section of this line is closed off by a pair of short circuits the result is a resonant section of line which can be excited by connecting a signal between the centres of the two sides of the slot as shown in Fig. 5.24(a). The resulting standing wave in the slot can be thought of as the superposition of a pair of equal and opposite travelling waves passing through the slot regarded as a short length of rectangular waveguide. It follows that a slot excited in this manner may be

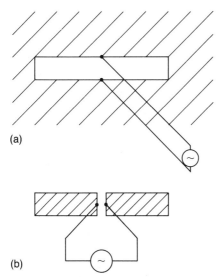

(a)

(b)

Fig. 5.24 (a) A slot antenna and (b) its complementary dipole.

expected to act as an antenna which radiates on both sides of the conducting sheet in which it is cut.

Booker's principle (Jordan *et al.*, 1968) states that the radiation from a slot antenna is the same as that from a dipole antenna which would just fill the slot (Fig. 5.24(b)) with the electric and magnetic fields interchanged. This dipole is called the complementary dipole. It can be shown that the radiation resistances of the slot antenna and its complementary dipole are related by

$$R_s R_d = Z_0^2/4. \tag{5.91}$$

We do not usually require the slot to radiate on both sides of the plane. A single-sided antenna can be made by backing the slot with a resonant cavity as shown in Fig. 5.25 (resonant cavities are discussed in Chapter 7). Because the slot now radiates only on one side and the resonant cavity presents an open circuit on the other side the radiation resistance is double that of the open slot. The slot can be excited by exciting the cavity with a probe or by connecting the two wires of a transmission line to opposite sides of the slot. Thus the radiation resistance of a cavity-backed half-wave slot radiator is

$$R_s = \frac{2 \times (377)^2}{4 \times 73} = 973 \,\Omega. \tag{5.92}$$

This impedance is too high for it to be easy to match the antenna to a coaxial cable. This problem can be solved by connecting the cable across the slot close to one end where the voltage is much lower. This arrangement

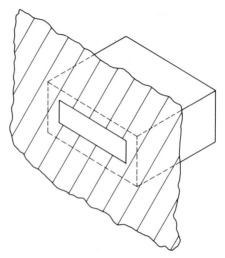

Fig. 5.25 Arrangement of a cavity-backed slot antenna.

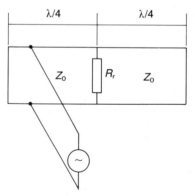

Fig. 5.26 An off-centre fed slot antenna.

has the equivalent circuit shown in Fig. 5.26 where Z_0 is the characteristic impedance of the slot line.

Another way of exciting a slot antenna is to cut it into the wall of a waveguide, as shown in Fig. 5.27(a), so that it intercepts the current circulating in the wall. An array of slots (Fig. 5.27(b)) can act as a compact highly directional antenna. The slots are carried over on to the broad walls of the guide to make them long enough to resonate within the waveguide band. If they are arranged at half guide-wavelength intervals then angling them as shown ensures that they are excited in phase with each other so that the array radiates in a broadside pattern.

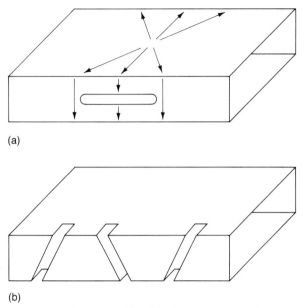

Fig. 5.27 Slot antennas in waveguide: (a) showing how a slot is excited by the currents flowing in the walls of the guide, and (b) a multi-slot array.

5.11 PHASED ARRAY ANTENNAS

From what has been said earlier about the radiation patterns of dipole arrays it will be clear that the radiation pattern can be altered by varying the relative phases of the array elements. This idea can be extended to cover arrays of other types of radiator. By making a two-dimensional array of elements it is possible, in theory, to synthesize any desired radiation pattern within the limits imposed by the overall size of the array. One reason why this idea is very attractive is that it allows the beam of the antenna to be steered electronically instead of mechanically. Mechanical scanning is limited by the inertia of the antenna. Electronic scanning offers the possibility of steering the beam almost instantaneously. This is clearly very attractive for military radar systems because it allows several fast-moving targets to be tracked simultaneously.

The disadvantage of phased arrays is that they must be very complex if they are to achieve good resolution. A typical array might have 100×100 elements. It could scan to a little over $45°$ either side of the normal to the array. Beyond this point the resolution deteriorates because the effective aperture of the antenna is smaller. Thus to achieve full coverage of the sky without moving the antenna five arrays arranged in a truncated four-sided pyramid are required.

Two different implementations of phased arrays are possible as shown in

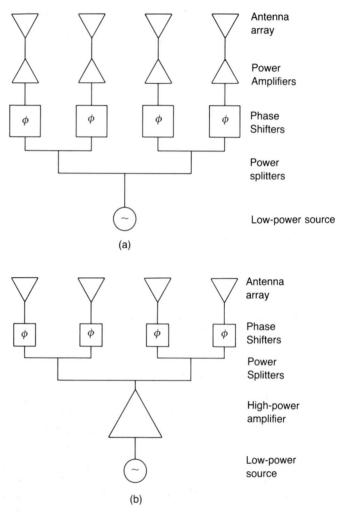

Fig. 5.28 Phased array antenna block diagrams: (a) system with low-power phase-shifters and distributed power amplifiers, and (b) system with a single high-power amplifier and high-power phased-shifter.

Figs. 5.28(a) and (b). The first employs a low-power signal source and carries out the power splitting and phase shifting at low power. There are then as many power amplifiers as antennas and each can be of quite low power. For example a 100×100 array of 10 W amplifiers would radiate 100 kW. The second arrangement uses a single high-power amplifier, probably a microwave tube, and carries out the splitting and phase shifting at higher power levels. The question of which of these two approaches is to be preferred is a complex one involving factors such as cost and reliability which we cannot pursue here. For further information see Rudge *et al.* (1982–3).

5.12 CONCLUSION

In this chapter we have seen how the alternating current in an electric or magnetic dipole can produce a radiating spherical TEM wave. The magnetic vector potential was introduced as a way of dealing with this kind of problem. It was also noted that the finite time taken for a wave to travel from one point to another means that the phase of the wave must be related to that of the source at an earlier time. This can be done by using the concept of retarded potentials. The power radiated from a dipole appears to the source to be dissipated in a resistive load whilst the energy stored in the near field provides a reactive component of impedance. For maximum power transfer the radiation resistance of the antenna must equal the source impedance. The reciprocity theorem was used to link the concept of the gain of a transmitting antenna to the effective area of a receiving antenna. The properties of a small magnetic dipole were deduced from those for a small electric dipole by using the principle of duality.

The radiation from apertures such as waveguide horns and dish antennas was considered based on Huygens' method of secondary sources. It was shown that the radiation pattern of these antennas is the Fourier transform of the illumination of the aperture so that narrow beams require large-aperture antennas for their generation.

Other practical antennas were discussed including the half-wave dipole, quarter-wave monopole and the slot antenna regarded as a complementary wire dipole. The use of arrays of antennas to provide particular radiation characteristics was introduced and the subject of active phased arrays was briefly touched on.

EXERCISES

5.1 Compute the effective areas of antennas having 20 dB gain at frequencies of 500 MHz, 4 GHz and 35 GHz.

5.2 Calculate the radiation resistance of a 100-turn coil wound on a non-magnetic former 10 mm in diameter at a frequency of 1 MHz.

5.3 Plot the polar radiation diagrams and calculate the maximum directivities for antennas comprising two parallel quarter-wave dipoles with the following spacings and phase differences: (a) $\lambda/4$, zero, (b) $\lambda/4$, 180° and (c) λ, 180°.

5.4 Calculate the directivities of endfire arrays of two, four and eight quarter-wave dipoles at frequencies 10% above and below the frequency at which they are a half-wavelength apart.

5.5 Plot polar diagrams for the radiation from a uniformly illuminated slot 100 mm wide at frequencies of 5 GHz, 10 GHz and 60 GHz.

5.6 Find the correct point for connecting the feed to a cavity-backed slot antenna as illustrated in Fig. 5.26 if the radiation resistance is 1000 Ω, the characteristic impedance of the slot is 100 Ω and the source impedance is 50 Ω.

Coupling between wave-guiding systems

6

6.1 INTRODUCTION

We have already seen that transmission lines and waveguides can support a variety of modes of propagation. Many important microwave devices incorporate transitions between regions with different propagation characteristics. In this chapter we shall examine the basic principles underlying coupling between these different regions and show how they are related to the theory and design of microwave components. We shall see that the coupling can often be thought of in terms of either the electric field or the magnetic field. For convenience all kinds of waveguiding structures will be referred to as waveguides. Such a huge variety of examples of coupling exists that it will be necessary to study only a representative sample and concentrate on the fundamental principles in operation. The various devices are usually represented by equivalent circuits and these are the basis of design calculations. Many of the equivalent circuits have been derived empirically and throw no light on the basic physical principles at work. For this reason a detailed discussion is not given here and the reader is referred to the books by Marcuvitz (1986), and Edwards (1981) for further details.

6.2 DISCONTINUITIES

All couplings between waveguides involve some kind of discontinuity in the waveguiding structure. It is therefore useful to begin by establishing what the effects of these discontinuities are.

Figure 6.1(a) shows a diaphragm which partially obstructs the width of a waveguide. At the plane of the diaphragm the fields must satisfy the boundary conditions given in Section 1.9. The field distribution at the plane of the diaphragm is a compressed TE_{01} mode as shown in Fig. 6.1(b) with $E_x = 0$ outside the gap. The field in the waveguide adjacent to the diaphragm must match that at the diaphragm in every respect. The field of

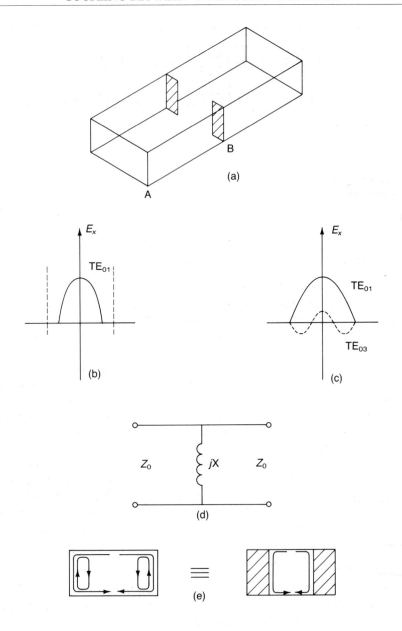

Fig. 6.1 An inductive waveguide iris: (a) general arrangement, (b) the field within the iris, (c) superposition of the waveguide TE_{01} mode and higher-order modes to satisfy the boundary conditions, (d) equivalent circuit, and (e) showing how the current in the plane of the iris can be represented by the superposition of inductive current loops on the current flow in the walls of the waveguide.

the incident TE_{01} wave shown in Fig. 6.1(c) plainly does not obey this requirement. In order to match the boundary conditions correctly we recall that an infinite set of higher-order modes can exist in the guide. The boundary conditions can thus be satisfied by a superposition of all these modes with appropriate amplitudes. Figure 6.1(c) shows the TE_{03} mode in the waveguide. Clearly the addition of the field of this mode to that of the TE_{01} mode produces a field distribution which is more nearly the same as that at B. In this case the amplitude of the TE_{02} would be zero because of the symmetry of the problem.

We have, therefore, established the basic principle that any discontinuity in a waveguide results in the transfer of energy to higher-order modes. Now the working frequency bands of waveguides are normally chosen so that the higher-order modes are cut off. The power coupled into them, therefore, does not propagate but is stored in evanescent waves close to the discontinuity. For many purposes the effect of the discontinuity can be represented as a lumped reactance at its plane. In the case of the diaphragm shown in Fig. 6.1(a) the effect of the discontinuity is to alter the paths of the conduction currents in the walls of the waveguide. Figure 6.1(e) shows how the effect of the diaphragm in altering the current paths can be represented as the superposition of a pair of current loops on the current flow in the unmodified waveguide. The diaphragm therefore introduces a shunt inductance and the discontinuity can be represented by the equivalent circuit shown in Fig. 6.1(d).

If the diaphragm changed the height of the guide, instead of its width, as shown in Fig. 6.2(a) a different set of higher-order modes would be excited. Once again these would normally be cut off. The discontinuity affects the electric field distribution so the shunt reactance is capacitive. An alternative way of introducing a shunt capacitance is to put a screw in the broad wall of the waveguide as shown in Fig. 6.2(b). This arrangement is useful for matching purposes because the capacitance is easily adjusted. An important special case arises when the diaphragm constricts both the height and the width of the guide as shown in Fig. 6.2(c). This arrangement presents a parallel combination of inductance and capacitance to the incident wave. The equivalent circuit is shown in Fig. 6.2(d). The arrangement is known as an iris and its resonant properties make it an important component in certain types of waveguide filter.

A comprehensive discussion of discontinuities in microstrip is given by Edwards (1981). Two are considered here to illustrate other ways in which discontinuities can be represented by equivalent circuits. We have already noted that it is easier to make open circuits than short circuits in microstrip. Thus matching stubs are normally open circuited at their free ends. The calculation of the lengths of such stubs is an important part of circuit design. Figure 6.3(a) shows the electric field in the region of an open circuit termination. It is clear that the field does not stop abruptly at the end of the

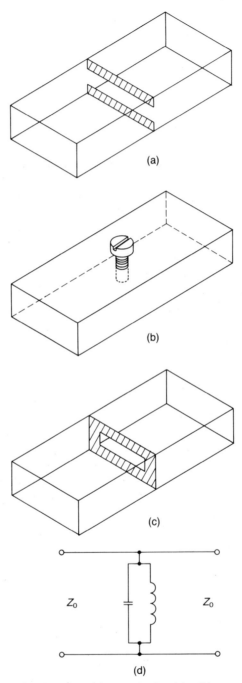

Fig. 6.2 Waveguide obstacles: (a) capacitative iris, (b) capacitative tuning screw, (c) resonant iris, and (d) equivalent circuit for the resonant iris.

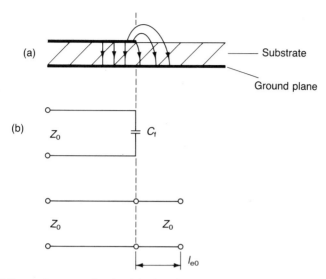

Fig. 6.3 Microstrip open circuit: (a) electric fringing field, (b) equivalent capacitance, and (c) equivalent transmission line.

line but fringes into the region beyond. This fringing field can be allowed for by a capacitor as shown in Fig. 6.3(b). The capacitance can be calculated accurately enough from a static field solution. Edwards quotes a formula for C_f.

An alternative way of looking at this effect is to consider that a short length of open-circuited transmission line will look like a capacitance. Thus the fringing can be represented by an additional length l_{e0} of the same

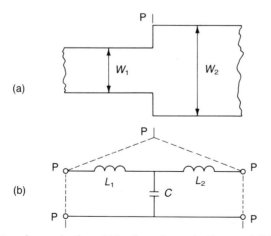

Fig. 6.4 (a) Step change in the width of a microstrip line, and (b) an equivalent circuit for the step.

transmission line as shown in Fig. 6.3(c). In design calculations, therefore, the line should be shorter than the theoretical length by l_{e0}.

As a final example consider the step change in the width of a microstrip line shown in Fig. 6.4(a). It has been suggested that the effects of the discontinuity can be represented by the equivalent circuit of Fig. 6.4(b). Here there is no clear physical link between the form of the equivalent circuit and the field patterns occurring at the junction.

The three examples given illustrate between them the various approaches which are used to produce equivalent circuits for microwave components. The first two are based upon a physical understanding of what is happening at the discontinuity. The last is a network whose elements are adjusted empirically to give useful results.

6.3 BROADBAND MATCHING TECHNIQUES

Mismatches in transmission lines can readily be matched at spot frequencies by arranging other mismatches to provide a reflected wave equal and opposite to the one to be eliminated. The matching elements take the form of shunt stubs in coaxial line and microstrip and inductive or capacitive irises in waveguide. A full description of the use of the Smith chart to design matching networks of this kind can be found in Dunlop and Smith (1984).

For many purposes this kind of narrow-band matching is not satisfactory and some way of producing a broad-band match is needed. To show how this can be achieved consider Fig. 6.5 which shows a transmission line with a matched termination which has four equally spaced reflecting objects with reflection coefficients ϱ_1 and ϱ_2 as shown. The electrical length of the line between the objects is ϕ. We will assume that the reflection coefficients are small so that the amplitude of the incident wave is hardly affected by the reflections. Now suppose that D is the obstacle to be matched and A, B and C are the matching elements. The apparent reflection coefficient at A is

$$\varrho = \varrho_1 + \varrho_2 e^{-2j\phi} + \varrho_2 e^{-4j\phi} + \varrho_1 e^{-6j\phi}$$
$$= 2\varrho_1 e^{-3j\phi} \cos 3\phi + 2\varrho_2 e^{-3j\phi} \cos \phi. \tag{6.1}$$

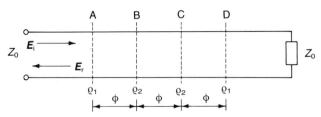

Fig. 6.5 Arrangement of four regularly spaced discontinuities on a transmission line.

In practical terms it is usually the magnitude of the reflection which is significant, that is

$$|\varrho| = |8\varrho_1 \cos^3 \phi + 2(\varrho_2 - 3\varrho_1) \cos \phi| \qquad (6.2)$$

since $\qquad \cos 3\phi = 4 \cos^3 \phi - 3 \cos \phi$.

Equation (6.2) may be written

$$|\varrho| = |8\varrho_1 x^3 + 2(\varrho_2 - 3\varrho_1)x|, \qquad (6.3)$$

where, for convenience, we have written $x = \cos \phi$. Now ϕ is a function of frequency and, therefore, so is x. Equation (6.3) describes the variation of the reflection coefficient with frequency. Clearly the shape of this cubic curve can be changed by making different choices of ϱ_2.

One possibility is to make $\varrho_2 = 3\varrho_1$. The result is

$$|\varrho| = |8\varrho_1 x^3| \qquad (6.4)$$

shown as the continuous curve in Fig. 6.6. A polynomial for which only the coefficient of the highest power of x is non-zero is known as a maximally flat polynomial. It has the property that the highest possible number of derivatives of the function are zero at the origin and that the curve is therefore as flat as possible close to the origin. The match having this frequency dependence is called a maximally flat or Butterworth response. It is the best possible match over a narrow band using a given number of matching elements.

For a given maximum reflection coefficient ϱ_m the bandwidth is the distance between the two points on the curve at which $\varrho = \varrho_m$ so that, from (6.4)

$$B_B = 2x = 2\left(\frac{\varrho_m}{8\varrho_1}\right)^{\frac{1}{3}} = \left(\frac{\varrho_m}{\varrho_1}\right)^{\frac{1}{3}}. \qquad (6.5)$$

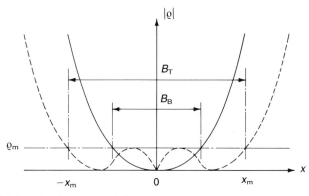

Fig. 6.6 Third-order Butterworth (maximally flat) and Tchebychev (equal-ripple) reflection characteristics.

This demonstrates that, as one would expect, it is possible to achieve a greater bandwidth by accepting a poorer match.

For broad-band matching it is better to sacrifice some of the excellence of the match close to the origin in order to gain greater bandwidth. The broken curve in Fig. 6.6 shows such a possibility. There are, of course, an infinite number of possible polynomials but of special interest are the Tchebychev polynomials. The first six polynominals are listed in Table 6.1.

Table 6.1 Tchebychev polynomials

n	$T_n(x)$
0	1
1	x
2	$2x^2 - 1$
3	$4x^3 - 3x$
4	$8x^4 - 8x^2 + 1$
5	$16x^5 - 20x^3 + 5x$

These polynomials have the property that over the range $-1 < x < 1$ the magnitude is always less than or equal to 1. Of all possible polynomials of a given order and specified in-band ripple the Tchebychev polynomial has the broadest bandwidth.

To design a broad-band match using Tchebychev polynomials we introduce the scaling factors ϱ_m and x_m. The third-order polynomial can then be written

$$\left(\frac{\varrho}{\varrho_m}\right) = 4\left(\frac{x}{x_m}\right)^3 - 3\left(\frac{x}{x_m}\right). \tag{6.6}$$

The response curve has an in-band ripple equal to ϱ_m and the band edges are at $x = \pm x_m$ as shown in Fig. 6.6.

Equating coefficients of x^3 in (6.3) and (6.6) we obtain

$$4\varrho_m/x_m^3 = 8\varrho_1 \tag{6.7}$$

so that

$$x_m^3 = \varrho_m/2\varrho_1. \tag{6.8}$$

Now ϱ_1 is the reflection coefficient of the termination to be matched and ϱ_m is the maximum acceptable in-band match. Therefore, from (6.8), the bandwidth which can be achieved in this way using three matching elements is

$$B_T = 2x_m = 2\left(\frac{\varrho_m}{2\varrho_1}\right)^{\frac{1}{3}} = 2^{\frac{2}{3}}\left(\frac{\varrho_m}{\varrho_1}\right)^{\frac{1}{3}}. \tag{6.9}$$

Comparing (6.9) with (6.5) we see that the bandwidth of the Tchebychev match is wider than that of the Butterworth match by a factor of $2^{2/3} = 1.59$ when three matching elements are used.

To complete the design of the Tchebychev matching network we equate the coefficients of x in (6.3) and (6.6) to give

$$2(\varrho_2 - 3\varrho_1) = -3\varrho_m/x_m \qquad (6.10)$$

so that

$$\varrho_2 = 3\varrho_1 - 3\varrho_m/2x_m. \qquad (6.11)$$

Since ϱ_1, ϱ_m and x_m are all known ϱ_2 can be calculated.

Example

Design third-order Butterworth and Tchebychev matching networks to give a matched reflection coefficient of 0.001 for an obstacle which has a reflection coefficient of 0.01.

Solution

For a Butterworth match the value of ϱ_2 is 0.03 and the bandwidth is 0.47 from (6.5).

For a Tchebychev match $x_m = 0.368$, the bandwidth is 0.74 from (6.9) and the value of ϱ_2 is 0.0259 from (6.11).

It must be remembered that the theory developed above assumes that all the reflection coefficients are small. It can illustrate the principles of broadband matching but cannot be applied to the problem of designing matching networks for large mismatches. However methods have been developed for designing both Butterworth and Tchebychev matching networks for large mismatches. (Matthaei *et al.*, 1964)

6.4 COUPLING WITHOUT CHANGE OF MODE

Microwave circuits involve bends and junctions in the waveguides which introduce discontinuities into the system. Some of these junctions do not involve any change in the mode of propagation. These are discussed here and those in which a change of mode is involved are discussed in the next section.

One of the simplest waveguide components is the bend. Provided that the bend is gradual the mismatch is slight and no problems arise. This is normally the case with coaxial lines. Sometimes, however, it is necessary to arrange a change of direction in rather a small space and the design of the bend then becomes important.

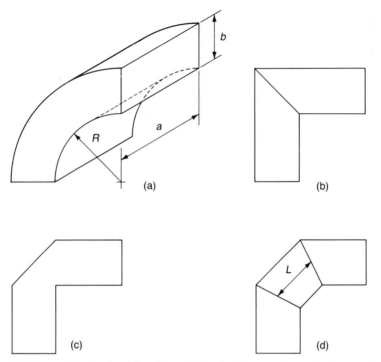

Fig. 6.7 Waveguide bends: (a) radiussed bend, (b) mitred bend, (c) chamfered bend, and (d) double mitred bend.

Waveguide bends are often made in the form of circular arcs as shown in Fig. 6.7(a). The direction of the bend can be either in the *E* plane or the *H* plane. Southworth (see Harvey, 1963) has shown that the minimum radius for a satisfactory match is 1.5 times the width of the guide in the plane of the bend. Thus for the bend shown in Fig. 6.7(a) the minimum value of *R* is 1.5*b*. Very accurate manufacturing is necessary to avoid reflections and tight bends are usually electro-formed. More gentle bends can be made by bending a straight waveguide using special equipment.

Radiussed waveguide bends are expensive so it is sometimes better to use fabricated mitred bends instead. Figure 6.7(b) shows a simple right-angle bend. Clearly the field patterns within the bend do not match those in the connecting waveguides so an appreciable reactive mismatch can be expected. An improvement (Fig. 6.7(c)) would be to chamfer the outside of the bend to make the effective height of the waveguide the same as that of the connecting guides. Better still if space allows is to use a double-mitred bend as shown in Fig. 6.7d. If the separation *L* between the two joints is made equal to an odd number of half wavelengths then the mis-matches caused by the junctions cancel each other out at one frequency

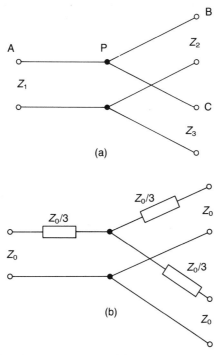

(a)

(b)

Fig. 6.8 Transmission line tee junctions: (a) simple tee, and (b) a matched tee.

and produce near cancellation over a useful band of frequencies. Similar techniques are used in microstrip.

At low frequencies we are accustomed to join three or more wires together at a point without thinking about the details of what we are doing. At microwave frequencies more care is needed. Figure 6.8(a) shows three transmission lines meeting at a point. If the characteristic impedances of the lines are as shown, A is the input port, and B and C are matched then the impedance presented at P is Z_2 in parallel with Z_3. Very often the impedances of the three lines will be the same and there is then clearly an appreciable mismatch at P. One solution to the problem is to incorporate a matching network of some kind between A and P. This approach leaves B and C incorrectly matched. If it is essential that all the ports are matched then the solution is to add series resistors as shown in Fig. 6.8(b). A little thought shows that this arrangement is correctly matched at all the ports but at the expense of dissipating half the input power within the device.

The networks shown in Fig. 6.8 are idealized, in practice we must take account of the discontinuity in the lines at the junction. Figure 6.9(a) shows a microstrip T junction. By analogy with the step change in width shown in Fig. 6.4(a) we expect that the equivalent circuit will need to have a shunt

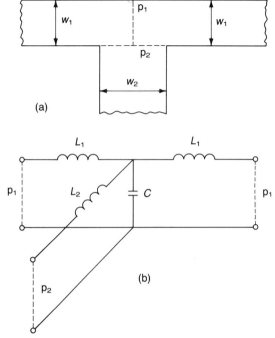

Fig. 6.9 Microstrip tee junction: (a) arrangement of the junction, and (b) equivalent circuit.

capacitance at the junction and series inductances in the three arms. The result is shown in Fig. 6.9(b). One important matter is the definition of the planes on the microstrip which are represented by the terminals of the equivalent circuit. For the main line it is natural to select the plane of symmetry (p_1) whilst for the side arm a plane such as p_2 might be used. It turns out that at frequencies above about 1 GHz it is necessary to add a transformer in the side arm whose turns ratio is a function of frequency. Similar equivalent circuits apply to other types of waveguide.

As a final example of this kind of coupling consider the four-port waveguide junction shown in Fig. 6.10(a). This arrangement combines an *E*-plane tee with an *H*-plane tee and is known as a hybrid tee or, sometimes, a magic tee. It has the property that any power fed in at port 1 is divided equally between ports 2 and 3, and any power fed in at port 4 is similarly divided. There is a very high isolation between ports 1 and 4. The reason why it works in this fashion can be understood from sketches of the fields for the two possible excitations. Figure 6.10(b) shows the field patterns when port 1 is the input. The signals emerge from ports 2 and 3 in phase with each other. The electric field at the entrance to the fourth arm is attempting to set up a TM_{11} wave, but this is cut off so the result is a

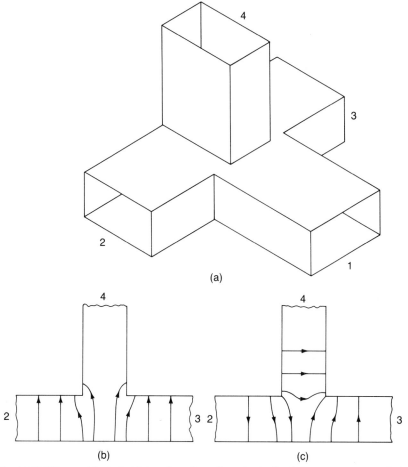

Fig. 6.10 Waveguide hybrid tee junction ('magic tee'): (a) arrangement of the junction, (b) electric fields when the input is at port 1, and (c) electric fields when the input is at port 4.

reactive loading of the junction. In the same way Fig. 6.10(c) shows the field patterns when port 4 is the input. This time the signals emerging from ports 2 and 3 are in antiphase and the wave excited in the remaining arm is the cut-off TE_{02} mode. The reactive elements of the junction are matched by a capacitive boss at its centre. It is difficult to match this device over a wide band.

6.5 COUPLING WITH CHANGE OF MODE

In the previous section we considered examples of coupling between waveguides which were propagating the same mode. Very often it is necessary

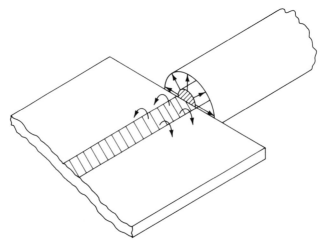

Fig. 6.11 Junction between coaxial line and microstrip.

to couple power from one mode to another. Examples of this are transitions from coaxial line to waveguide and to microstrip and from one waveguide mode to another.

The most straightforward of these is the coupling from coaxial line to microstrip. This is of great practical importance because most low-power microwave circuits are realized in microstrip whilst the external connections between them are normally made using coaxial lines. Figure 6.11 shows a junction between these two types of line. The standard figure for the characteristic impedance of coaxial line is $50\,\Omega$ and there is no problem in designing a microstrip line having the same impedance. But, though that is a necessary condition for a match, it is not sufficient. The reason for this is clear from Fig. 6.11 which shows the difference in the electric field patterns of the two modes. It is therefore necessary to provide some matching elements to compensate for the reactance of the junction.

In many other cases there is a very poor match between the field patterns of the two modes which are to be coupled. When that is so the approach is to try to find some way of matching the field patterns as nearly as possible. In that way a high proportion of the power is coupled into the desired mode and relatively little into other modes. An example of this is the transition from the TE_{01} rectangular waveguide mode to the TM_{01} circular waveguide mode required to make the rotating waveguide joints used to feed power to radar antennas. Figure 6.12(a) shows the field patterns in the two guides. It is immediately evident that the magnetic field patterns are similar if the guides are arranged at right angles to each other as shown. The electric fields are not so well matched because the removal of part of the broad wall of the rectangular guide means that the field lines have to be

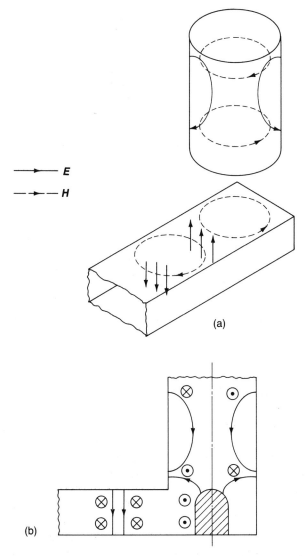

Fig. 6.12 TE_{10} rectangular waveguide to TM_{01} circular waveguide junction: (a) arrangements of the fields in the waveguides, and (b) a cross-sectional view of a 'door-knob' transition showing the field patterns.

radically redistributed. The solution is to put a boss (A) at the centre of the transition as shown in Fig. 6.12(b). Additional irises may be needed to match the transition completely over a band of frequencies. This transition is known as a 'door-knob' transition.

As a final example we will consider the problems of making a good

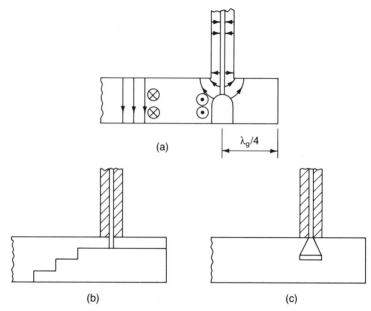

Fig. 6.13 Coaxial line to waveguide transitions: (a) door-knob transition, (b) ridge-waveguide transition, and (c) probe transition.

$$\frac{2b}{a} \frac{\lambda_g}{\lambda_0} Z_0$$

broad-band match from coaxial line to waveguide. Standard coaxial line has a characteristic impedance of $50\,\Omega$. The characteristic impedance of rectangular waveguide is given by (2.51). Standard waveguides have an aspect ratio (a/b) of about $2:1$ and are used over a range of λ_g/λ_0 of about 1.7 to 1.2. The impedance to be matched therefore ranges from $630\,\Omega$ to $440\,\Omega$, roughly ten times that of the coaxial line, and devices for this purpose are therefore often called 'coaxial-to-waveguide transformers'.

Consideration of the field patterns in the two guides suggests that a door-knob transition similar to that shown in Fig. 6.12(b) might be used. The field patterns arising are shown in Fig. 6.13(a). A quarter-wave length of waveguide behind the transition is used to make the shunt impedance of the end of the waveguide an open circuit at the transition. It can be seen that the field patterns are not very well matched.

One of the difficulties of the door-knob transition is that it has to cope with the change from a low-voltage, high-current, wave in the coaxial line to a high-voltage, low-current, wave in the waveguide. The existence of a direct current path makes this difficult. Two solutions to this problem suggest themselves: one is to make use of a lower impedance waveguide, the other is to avoid a direct current path.

To reduce the waveguide impedance we must reduce its height (see (2.51)). But the reduction to something like one tenth of the height of

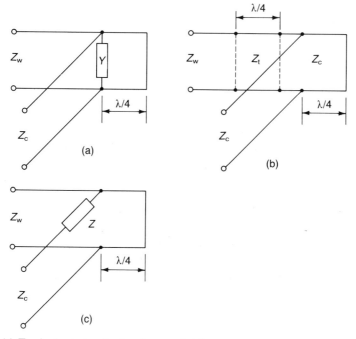

Fig. 6.14 Equivalent circuits for the coaxial line to waveguide transitions shown in Fig. 6.13.

standard waveguide presents new problems for the transformation between the two. A better solution is to use ridge waveguide which can have an impedance of 50 Ω without needing such narrow gaps. Figure 6.13(b) shows a transition using a ridge waveguide with a stepped impedance transformer to standard waveguide and, once again, a quarter-wave section of short-circuited line behind it.

The third approach is to couple into the waveguide using displacement current rather than conduction current. Figure 6.13(c) shows how this is achieved. The free end of the centre conductor is, in effect, an antenna radiating into the waveguide. The use of an expanded end to it provides for the necessary impedance transformation. It also has the effect of providing a good match between the field patterns as can be seen from Fig. 6.13(c). These two statements are really just saying the same thing, one in circuit terms and the other in field terms.

The three transitions described above can be represented by the equivalent circuits shown in Figs. 6.14(a), (b) and (c), respectively. The first is unsatisfactory and is seldom used. The second works well but is bulky. The third can be made to have a good broadband match and has the advantage of being very compact. Many other kinds of transition have been invented, some of which are in common use. The three described here have been included to illustrate the principles involved in making a mode transition of this kind. Further details can be found in Harvey (1963) and Marcuvitz (1986).

6.6 COUPLING BY APERTURES

So far we have discussed cases where it is desired to transfer all the power flowing in one waveguide into another so that the guides are strongly coupled; sometimes this is not required. Consider, for example, the situation shown in Fig. 6.15. Two rectangular waveguides are placed side by side with a small hole in their common wall. It is reasonable to suppose that some of the power flowing in waveguide 1 will pass through the hole and excite waves in waveguide 2 as shown. Such a device has possible uses as a way of sampling the signal in waveguide 1. It is interesting, therefore, to consider the theory of coupling by small holes.

The current flowing in the wall of the waveguide is diverted by the presence of the hole as shown in Fig. 6.16. Provided that the hole is small,

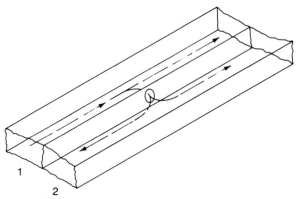

Fig. 6.15 Coupling between waveguides by a small hole in a common wall.

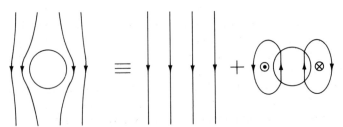

Fig. 6.16 Representation of the current flow around a hole in a waveguide wall by the superposition of the field of a magnetic dipole on a uniform field.

Fig. 6.17 The waves radiated into two waveguides by a magnetic dipole in their common wall.

so that the current patterns away from the hole are undisturbed and the hole itself is small compared with the free-space wavelength, we can think of its effect as the superposition of a small current loop on the undisturbed current flow. This current loop generates an alternating magnetic field and so acts as a magnetic dipole antenna which radiates power into both wave-guides. The radiated power excites four waves as shown in Fig. 6.17 besides exciting cut-off higher-order modes. S_{1+} must be roughly in antiphase with S_0 because the hole is extracting some power from the incident wave. The backward wave S_{1-} shows that the presence of the hole introduces a mis-match into guide 1. The signals S_{2+} and S_{2-} provide samples of S_0 which can be fed to a detector or power meter. The whole system is linear so the sampled power is proportional to the incident power.

When the two guides have the broad wall as their common boundary the situation is a little more complicated. This is because there are now two components of the tangential magnetic field and one of the normal electric field. An argument similar to the one in the previous paragraph can be used. Figure 6.18(a) shows how the fringing of the normal electric field through a small hole can be represented by the superposition of a small electric dipole on the unperturbed field. Similarly, Fig. 6.18(b) shows the representation of the fringing of magnetic field lines through a small hole by a small magnetic dipole. The latter is an alternative way of looking at

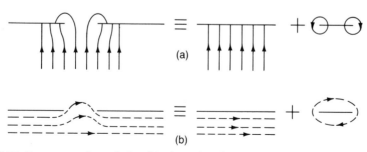

Fig. 6.18 Representation of the fringing electric and magnetic fields around a hole in a conducting wall by the superposition of a dipole field on a uniform field: (a) electric fringing field, and (b) magnetic fringing field.

the situation shown in Fig. 6.16. A little thought shows that the forward and backward waves radiated into a guide by an electric dipole are in phase with each other whilst those due to a magnetic dipole are in antiphase.

The dipole moments are evidently proportional to the excitation and may be expected to depend upon the size and shape of the hole. The dependence of the dipole moment on the size of the hole is expressed as its polarizability, so that

$$p = \varepsilon_0 \alpha_e |E_i| \qquad \text{and} \qquad j = \mu_0 \alpha_m |H_i|, \tag{6.12}$$

where p and j are the electric and magnetic dipole moments, α_e and α_m are the electric and magnetic polarizabilities and $|E_i|$ and $|H_i|$ are the exciting fields. For a circular hole of radius r the polarizabilities are given by

$$\alpha_e = \tfrac{2}{3} r^3 \qquad \text{and} \qquad \alpha_m = \tfrac{4}{3} r^3. \tag{6.13}$$

The mathematical theory of coupling by small holes is rather complicated so it will not be reproduced here. Details can be found in Collin (1966). In practice the design of devices which use coupling holes generally proceeds of the basis of measured parameters (see Harvey, 1963).

6.7 EFFECT OF HOLES IN SCREENS ON SCREENING EFFECTIVENESS

In Chapter 4 we saw that a conducting enclosure can be a very effective screen against electromagnetic interference provided that there are no holes in it. We are now in a position to consider how the presence of holes might affect the screening effectiveness of an enclosure.

Figure 6.19(a) shows a plane TEM wave incident normally on a conducting sheet with a square hole in it. The thickness of the sheet is d and

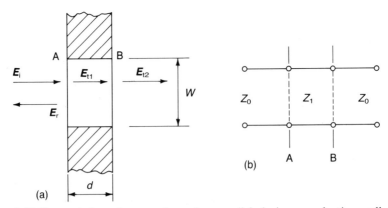

Fig. 6.19 Transmission of waves through a small hole in a conducting wall: (a) general arrangement, and (b) transmission-line equivalent circuit.

the width of the hole is W. Figure 6.19(b) shows the equivalent circuit for this problem. The hole normally has a width much less than the free-space wavelength so it behaves as a strongly cut-off waveguide and the signal is heavily attenuated as it passes through. The impedance mismatches at A and B are large giving rise to strong reflections and the possibility of multiple reflections if the attenuation of the hole is not very great. Thus the screening effectiveness of the sheet with the hole in it can be described by the equation

$$S = A_a + R_a + B_a \qquad (6.14)$$

where A_a is the attenuation of the signal as it passes through the aperture, R_a is the reflection of the signal at the transition from the aperture to free space, and B_a is a term to account for multiple reflections.

The propagation constant in a rectangular waveguide is given by (2.64) and, for the lowest TE mode $k_c = \pi/W$. Now, if $k_0 \ll k_c$, $k_g = jk_c$. The signal therefore propagates through the aperture as

$$E = E_0 \exp - (\pi z/W), \qquad (6.15)$$

so that

$$A_a = -20 \log_{10} [\exp - (\pi d/W)]$$
$$= 27.3(d/W) \, dB. \qquad (6.16)$$

The wave impedance of the wave in the aperture is

$$Z_1 = \frac{E_x}{H_y} = \frac{\lambda_g}{\lambda_0} Z_0$$

$$= -\frac{2jW}{\lambda_0} Z_0 \qquad (6.17)$$

from (2.45) and (2.46). This impedance is typically much less than the wave impedance of free space so there is a substantial mismatch. Assuming that the mismatch of impedances dominates so that the effects of higher-order modes can be neglected we have, from (1.85)

$$\frac{E_{t1}}{E_i} = \frac{2Z_1}{Z_0 + Z_1} \simeq -\frac{2jW}{\lambda_0} \qquad (6.18)$$

so that

$$R_a \simeq -20 \log_{10} (2W/\lambda_0). \qquad (6.19)$$

To find the effects of multiple reflections consider Fig. 6.20. If the amplitude of the wave entering the hole at A is E_0 then in the absence of multiple reflections the amplitude of the wave arriving at B is

$$E_f = E_0 e^{-\alpha d}. \qquad (6.20)$$

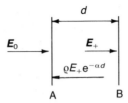

Fig. 6.20 Multiple reflection of waves within an aperture in a conducting sheet.

Now the voltage-reflection coefficients at A and B for waves travelling within the hole are both

$$\varrho = \frac{Z_0 - Z_1}{Z_0 + Z_1} \tag{6.21}$$

from (1.84). So, if the wave is multiply reflected at A and B, we must have

$$E_+ = E_0 e^{-\alpha d} + \varrho^2 E_+ e^{-2\alpha d} \tag{6.22}$$

or

$$E_+ = \frac{E_0 e^{-\alpha d}}{(1 - \varrho^2 e^{-2\alpha d})}. \tag{6.23}$$

The bottom line of this expression represents the effects of multiple reflections so

$$B_a = 20 \log_{10} (1 - \varrho^2 e^{-2\alpha d}). \tag{6.24}$$

Now $Z_0 \gg Z_1$ so $\varrho \simeq 1$ and (6.24) can be written

$$B_a \simeq 20 \log_{10} (1 - 10^{-A_a/10}). \tag{6.25}$$

To see the implications of this theory in a practical case let us consider the effect of making a small hole in the aluminium screen whose screening effectiveness was estimated in Chapter 4.

Example

Estimate the leakage of signals through a square hole 5 mm × 5 mm in an aluminium sheet 0.2 mm thick over the frequency range 10 Hz to 10 MHz.

Solution

Substituting these figures into (6.16), (6.19) and (6.24) we get the figures in Table 6.2.

Comparison of these figures with those computed in the last chapter shows that even a small hole can produce a dramatic reduction in the screening effectiveness of an enclosure. As might be expected the degredation gets worse as the frequency increases.

Table 6.2

f	A(dB)	R(dB)	B(dB)	S(dB)
10 Hz	1.1	190	−13	178
100 Hz	1.1	170	−13	158
1 kHz	1.1	150	−13	138
10 kHz	1.1	130	−13	118
100 kHz	1.1	110	−13	98
1 MHz	1.1	90	−13	78
10 MHz	1.1	70	−13	58

So far we have only considered the case of a single square hole. If the hole had been some other shape the same general principles would apply but the figures would have been slightly different. When there are several holes it is necessary to include corrections for the proportion of the surface area occupied by the holes and for the coupling between them. In the limit of a wire mesh an additional correction is needed to take account of the skin effects in the wires. For further discussion of this subject see Keiser (1979).

6.8 WAVEGUIDE DIRECTIONAL COUPLERS

We have seen in Section 6.6 that it is possible to couple power from one waveguide to another through small holes in a common wall. By a suitable choice of the positions of those holes it is possible to make the coupling directional. Figure 6.21 shows schematically such an arrangement. In ideal coupler if a signal P_1 were injected at port 1 signals P_2 and P_3 would appear at ports 2 and 3 with nothing at port 4. Then the coupling of the coupler is defined as

$$C = 10 \log_{10} \left(\frac{P_1}{P_3}\right). \tag{6.26}$$

From the symmetry of the device the coupling would be the same if the power were injected at any other port with a cyclical permutation of the port numbers.

Fig. 6.21 Schematic diagram of a directional coupler.

In practice no directional coupler is perfect and some of the incident power at port 1 is coupled to port 4. The closeness of the coupler to the ideal is measured by its directivity defined by

$$D = 10 \log \left(\frac{P_4}{P_3}\right). \tag{6.27}$$

Notice carefully that the directivity is the ratio of the forward and backward powers in the second guide and not the ratio of the backward power in the second guide to the incident power in the first guide.

A device of this kind has many uses in microwave systems. It can be used to measure the forward and backward powers in the primary guide with very little disturbance to the quantities being measured provided that the coupling is weak enough.

To illustrate the way in which it is possible to make a directional coupler consider the two-hole side-wall coupler shown in Fig. 6.22. Let the electrical distance between the two holes be ϕ and let each radiate forward and backward waves with amplitudes E for an incident wave amplitude E_i. The coupling is magnetic and we recall that the forward and backward waves radiated from each hole are in antiphase. Taking the plane A as the reference plane the amplitude of the forward wave in the secondary guide is

$$E_3 = E + (Ee^{-j\phi})e^{j\phi} \tag{6.28}$$

whilst the backward wave is

$$E_4 = -E - (Ee^{-j\phi})e^{-j\phi}. \tag{6.29}$$

The forward-wave components add for all values of the electrical length. The backward wave components cancel each other out if the electrical length between the holes is a quarter wavelength. This condition only exists at certain frequencies. In between, the backward power will be non-zero and could be equal to the forward power. Note that if the coupler is

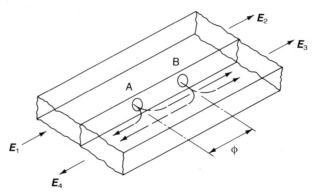

Fig. 6.22 Simple waveguide directional coupler.

symmetrical the backward wave will not quite be zero because the amplitude of the signal at B is less than that at A as a result of the transfer of some power at A from the first guide to the second. The bandwidth of the coupler can be increased by adding more holes to give either a Butterworth or a Tchebychev response (Levy, 1959). The coupling could equally well have been made through holes in a common broad wall though then the situation would have been complicated by the combination of electric and magnetic coupling.

Multi-hole directional couplers are commonly used where good directivity and a flat frequency resonse are required. The coupling is usually in the range 10 dB to 30 dB and a good instrumentation coupler has a directivity exceeding 40 dB throughout its band of operation.

Multi-hole directional couplers can be obtained with either three ports or four. The only difference is that in a three-port design the fourth port has been terminated by a built-in waveguide load. This can give a better match than if a separate external load were used. The reason why this is important can be seen from Fig. 6.21. Suppose that port 1 is connected to the signal generator whilst port 2 is connected to a load with a small reflection coefficient. Port 3 is terminated by a load and the power at port 4 is used to measure the reflection coefficient of the load in the main guide. The actual signal detected at port 4 will be made up of three components: first the backward power coupled from the main guide, second the sample of the forward power coupled to port 4 because of the finite directivity of the coupler, and third the reflection of the forward-coupled power by the imperfect load at port 3.

Example

A directional coupler has a coupling of 20 dB, directivity of 40 dB and an internal load whose return loss is 40 dB. Find the possible range of error if it is used to measure the return loss of a load whose actual return loss is 30 dB.

Solution

If the signal supplied by the signal generator is 1 mW (power levels are sometimes expressed in decibels referred to 1 milliwatt, written 'dBm', so this is 0 dBm) then the three components of the signal detected are:

power at port 2 coupled to port 4 = −50 dBm (10 nW);
power at port 1 coupled to port 4 = −60 dBm (1 nW); and
power at port 1 coupled to port 3 and reflected = −60 dBm (1 nW).

The power levels of the two smaller signals are each 1/10 of that of the signal to be measured so their fields are smaller by a factor of $1/\sqrt{10}$. The total signal level detected is

$$E = E_0 \pm \frac{2}{\sqrt{10}} E_0,$$

where the two signs represent the cases where the smaller waves are in phase or in antiphase with the signal to be measured. Thus the maximum and minimum signal levels are

$$1.632E_0 \quad \text{and} \quad 0.368E_0,$$

giving limits of $+4.25\,\text{dB}$ and $-8.68\,\text{dB}$.

Many other different configurations of coupler have been designed. One which is particularly useful because of its compactness is the cross-guide coupler shown in Fig. 6.23. The use of offset holes in a common broad wall means that each hole must be represented by one electric dipole and two magnetic dipoles. Because of the different phase relationships for these dipoles it is possible to make each hole have directional properties. As with the coupler shown in Fig. 6.22 the correct choice of hole separation can also give directional properties. Cross-guide couplers typically have couplings in the range 10 dB to 30 dB, but with directivities only a little better than 20 dB.

It is possible to make couplers with couplings as strong as 3 dB but it is correspondingly more difficult to achieve good directivity or a flat frequency response. The hybrid tee junction discussed earlier can be thought of as a 3 dB coupler. A pair of these couplers can be used as power combiners. Figure 6.24 shows one application. The incoming power is split at A into two branches one of which contains a variable phase shifter B. The signals are then recombined at C. As the phase shifter is adjusted the output power changes from zero up to the full input power. The couplers and the

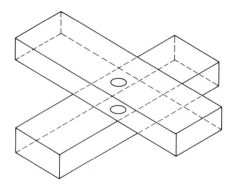

Fig. 6.23 Cross-guide directional coupler.

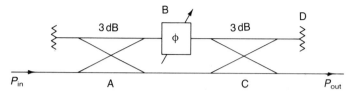

Fig. 6.24 Construction of a high-power attenuator using 3 dB couplers, a phase shifter and a high-power load.

phase shifter are lossless devices so the remaining power must end up in the load D. If D is a high-power load then this circuit can be used as a high-power attenuator.

6.9 DISTRIBUTED COUPLING

In the preceding section we have considered examples of coupling where the coupling is achieved by a number of discrete coupling elements. This is the kind of coupling normally encountered in waveguide devices. In microstrip and optical-fibre lines the coupling is usually distributed over a length of line. Figure 6.25 shows two parallel microstrip lines and the two possible ways in which they can be coupled together. In the first case (Fig. 6.25(a)) the signals on the lines are in phase with each other and the coupling is via the magnetic field. In the second case (Fig. 6.25(b)) the coupling is via the electric field and the signals on the lines are in antiphase. Notice carefully that in both these cases the signals on both lines are propagating into the paper as can be seen from the directions of the Poynting vectors. Note also that the two lines could be parallel tracks on a printed circuit board. Thus the theory given in this section applies also to cross-talk in high-speed

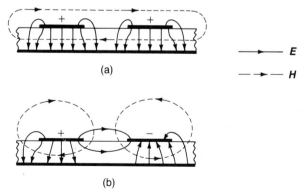

Fig. 6.25 Coupling between microstrip lines: (a) even-mode coupling, and (b) odd-mode coupling.

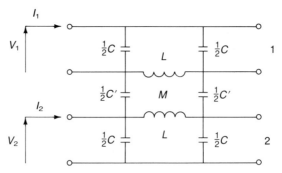

Fig. 6.26 Equivalent circuit for distributed coupling between transmission lines.

digital circuits. The coupled transmission lines can be represented by the equivalent circuit shown in Fig. 6.26. This shows pi sections of two identical transmission lines coupled by capacitance C' and mutual inductance M per unit length.

Using the same method of analysis as is used for single TEM lines (Carter, 1986) we can write down the set of equations which describe the coupled system.

$$\frac{\partial V_1}{\partial x} + L\frac{\partial I_1}{\partial t} = -M\frac{\partial I_2}{\partial t} \tag{6.30}$$

$$\frac{\partial V_2}{\partial x} + L\frac{\partial I_2}{\partial t} = -M\frac{\partial I_1}{\partial t} \tag{6.31}$$

$$\frac{\partial I_1}{\partial x} + (C + C')\frac{\partial V_1}{\partial t} = C'\frac{\partial V_2}{\partial t} \tag{6.32}$$

$$\frac{\partial I_2}{\partial x} + (C + C')\frac{\partial V_2}{\partial t} = C'\frac{\partial V_1}{\partial t} \tag{6.33}$$

These equations are an example of a set of coupled-mode equations. The left-hand side of each equation is essentially the equation describing an uncoupled mode whilst the right-hand sides provide the coupling terms. It is evident that if the coupling is removed (C' and M tending to zero) the equations reduce to those for the uncoupled lines. Coupled-mode theory (Pierce, 1954; Louisell, 1960) provides a valuable conceptual tool for studying problems involving coupling between wave-propagating systems.

Assuming propagation as $\exp j(\omega t - kx)$ we obtain

$$-jkV_1 + j\omega LI_1 = -j\omega MI_2 \tag{6.34}$$

$$-jkV_2 + j\omega LI_2 = -j\omega MI_1 \tag{6.35}$$

$$-jkI_1 + j\omega(C + C')V_1 = j\omega C'V_2 \tag{6.36}$$

$$-jkI_2 + j\omega(C + C')V_2 = j\omega C'V_1. \tag{6.37}$$

For the symmetric mode (Fig. 6.25(a)) $V_1 = V_2$ and $I_1 = I_2$ so that from (6.34) and (6.36) we obtain

$$-jkV_1 + j\omega(L + M)I_1 = 0 \tag{6.38}$$

$$-jkI_1 + j\omega CV_1 = 0, \tag{6.39}$$

whence

$$k^2 - \omega^2(L + M)C = 0 \tag{6.40}$$

so that the propagation constant for the symmetric mode is

$$k_+ = \pm\omega\sqrt{[(L + M)C]}. \tag{6.41}$$

By substitution into (6.39) we find that the characteristic impedance of this mode is

$$Z_+ = \sqrt{\left(\frac{L + M}{C}\right)}. \tag{6.42}$$

Notice that C' does not appear in either (6.41) or (6.42), showing that the coupling is pure magnetic.

For the antisymmetric mode $V_1 = -V_2$ and $I_1 = -I_2$. Analysis similar to that given above leads to the following expressions for the propagation constant and the characteristic impedance

$$k_- = \pm\omega\sqrt{[(L - M)(C + 2C')]} \tag{6.43}$$

$$Z_- = \sqrt{\left(\frac{L - M}{C + 2C'}\right)}. \tag{6.44}$$

The propagation constants of the two modes can be written

$$k_+ = \pm k_0\sqrt{\left(1 + \frac{M}{L}\right)} \tag{6.45}$$

$$k_- = \pm k_0\sqrt{\left[\left(1 - \frac{M}{L}\right)\left(1 + \frac{2C'}{C}\right)\right]}, \tag{6.46}$$

where $k_0 = \omega\sqrt{(LC)}$ is the propagation constant of the uncoupled lines. Note that k_+ is always greater than k_0. The product of k_+ and k_- is

$$k_+k_- = k_0^2\left[\left(1 - \frac{M^2}{L^2}\right)\left(1 + \frac{2C'}{C}\right)\right]^{\frac{1}{2}}. \tag{6.47}$$

For the special case of weak coupling the terms involving M and C' can be neglected and

$$k_+k_- \simeq k_0^2. \tag{6.48}$$

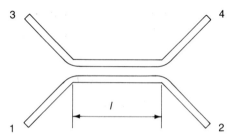

Fig. 6.27 Arrangement of a simple microstrip directional coupler.

Similarly, the product of the coupling impedances is

$$Z_+ Z_- = \sqrt{\left(\frac{L^2 - M^2}{C^2 + 2CC'} \right)} \tag{6.49}$$

and for weak coupling

$$Z_+ Z_- \simeq Z_0^2. \tag{6.50}$$

Now consider a microstrip directional coupler made by arranging two lines as shown in Fig. 6.27. An input signal at port 1 may be expected to excite both even and odd modes in the coupler. The analysis of this coupler proceeds by assuming forward and backward symmetric and antisymmetric waves in each arm (eight waves in all) and then applying the boundary conditions to obtain the amplitudes of each.

The analysis (Edwards, 1981) is tedious rather than difficult so it will not be reproduced here. The results may be expressed in terms of the signal amplitudes at the ports as

$$\frac{V_2}{V_1} = \frac{\sqrt{(1 - C^2)}}{\sqrt{(1 - C^2)} \cos \theta + j \sin \theta} \tag{6.51}$$

$$\frac{V_3}{V_1} = \frac{jC \sin \theta}{\sqrt{(1 - C^2)} \cos \theta + j \sin \theta} \tag{6.52}$$

$$V_4 = 0, \tag{6.53}$$

where $\theta = k_0 l$ (the difference between k_+ and k_- being neglected) and the coupling constant C is defined by

$$C = \frac{Z_+ - Z_-}{Z_+ + Z_-}. \tag{6.54}$$

The interesting thing about these results is that the coupled power appears at port 3 not at port 4. A coupler with this property is known as a contradirectional coupler as opposed to a co-directional coupler. Maximum coupling is obtained when the coupled lines are a quarter of a wavelength long and then the coupling is

$$20 \log_{10} \left(\frac{Z_+ - Z_-}{Z_+ + Z_-} \right). \qquad (6.55)$$

Because the electrical length of the lines varies with frequency the coupling is frequency dependent and the design is relatively narrow band.

It would appear from (6.53) that this kind of coupler must have infinite directivity. This is not the case because the difference between the propagation constants of the two modes has been neglected in the analysis. There are also, inevitably, reflections at the ends of the coupler where the modes are mismatched to the connecting lines. We have also assumed that the uncoupled modes can be regarded as pure TEM modes. In practice it is difficult to make a coupler of this kind with a directivity better than 15 dB.

The bandwidth of the coupler shown in Fig. 6.27 can be improved by the addition of lumped capacitors at its ends as shown in Fig. 6.28(a). The

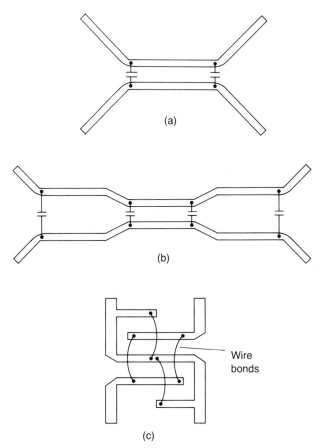

Fig. 6.28 Microstrip directional couplers: (a) narrow-band coupler, (b) broad-band coupler, and (c) Lange coupler.

capacitors have the effect of equalizing the phase velocities of the odd and even modes. Another approach (shown in Fig. 6.28(b)) is to make use of a number of sections to achieve a Tchebychev response.

An alternative type of microstrip directional coupler which is very commonly used is the Lange coupler shown in Fig. 6.28(c). With this arrangement couplers can be made with 10 dB coupling and 20 dB directivity over a frequency band of an octave. The design is largely empirical. For further details see Edwards (1981) and the references therein.

The discussion so far in this section has concentrated on microstrip couplers as examples of distributed directional couplers. It is possible for coupling between optical fibres to occur in a similar way. We have seen in Chapter 3 that there is an evanescent wave just outside the core of an optical fibre cable. If a second core is placed close enough to the first it will be coupled to the evanescent wave. The result is a 'leaky-wall' coupler which has distributed coupling and properties similar to the microstrip example discussed above. This coupling could result in crosstalk in bundles of optical fibres.

6.10 CONCLUSION

In this chapter attention has been focussed on phenomena involving coupling of modes of propagation. It has been shown that cut-off higher-order modes are excited at discontinuities in transmission lines and waveguides. These modes store energy in a region close to the discontinuity so they can be represented by lumped reactances. Coupling between different modes of propagation has been discussed in terms of the matching of their field patterns.

A discussion of the principles of coupling by small holes led to consideration of multi-hole directional couplers. Finally distributed directional couplers were introduced with microstrip and optical couplers as examples.

EXERCISES

6.1 Use the method described in Section 6.3 to investigate fourth-order Butterworth and Tchebychev matching using four matching elements with reflection coefficients ϱ_1, ϱ_2, ϱ_3 and ϱ_2. Derive expressions similar to (6.5) and (6.9) for the bandwidth in terms of the maximum mismatch. What is the ratio of the bandwidth of the fourth-order Tchebychev match to the fourth-order Butterworth match and the third-order Tchebychev match having the same maximum reflection?

6.2 Estimate the screening effectiveness of the enclosure described in the worked example on p. 146 if the hole is enlarged to 5 mm × 10 mm.

6.3 Design a transition at 6 GHz from WG14 waveguide to 50 Ω coaxial line by coupling the line to a 50 Ω section of reduced-height waveguide and using a single-stage quarter-wave transformer to match the reduced-height guide to the standard guide. Why is this technique not used?

6.4 Two microstrip lines each 2 mm wide are constructed with their centres 5 mm apart on an alumina substrate ($\varepsilon_r = 10$) 1.0 mm thick. Estimate the magnitudes of the equivalent circuit parameters shown in Fig. 6.26 and, hence, the even- and odd-mode characteristic impedances and the propagation constants at 900 MHz.

Electromagnetic resonators and filters

7.1 INTRODUCTION

At radio frequencies tuned circuits are used to provide frequency selectivity. Resonant elements are used for similar purposes at microwave and optical frequencies. They can take many forms but all can be modelled fairly accurately by one of the resonant circuits shown in Fig. 7.1. The impedance of the parallel resonant network (Fig. 7.1(a)) is given by

$$Z = \frac{R}{1 - j\dfrac{R}{\omega L}(1 - \omega^2 LC)}. \tag{7.1}$$

At microwave frequencies the individual circuit components have no real significance and it is usual to rewrite (7.1) in the form

$$Z = \frac{R}{1 + jQ\left(\dfrac{\omega}{\omega_0} - \dfrac{\omega_0}{\omega}\right)}, \tag{7.2}$$

where
$$\omega_0 = 1/\sqrt{(LC)} \tag{7.3}$$

and
$$Q = R/\omega_0 L = \omega_0 RC. \tag{7.4}$$

To understand what this expression means let us consider separately the amplitude and phase of Z. The amplitude is

$$|Z| = \frac{R}{\left[1 + Q^2\left(\dfrac{\omega}{\omega_0} - \dfrac{\omega_0}{\omega}\right)^2\right]^{\frac{1}{2}}}. \tag{7.5}$$

This expression clearly has a maximum value of $|Z| = R$ when the frequency-dependent term on the bottom line is zero. This occurs at the resonant frequency of the network given by (7.3). At all other frequencies $|Z| < R$.

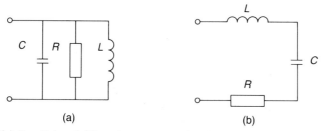

Fig. 7.1 (a) Parallel and (b) series resonant circuits.

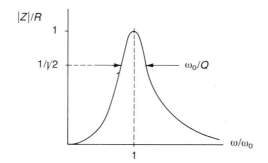

Fig. 7.2 Amplitude-response curve of a parallel resonant circuit.

Figure 7.2 shows how $|Z|$ varies with frequency. The width of the curve is determined by the parameter Q in (7.2). The relationship is usually expressed in terms of the width of the curve at the 3 dB points. If the circuit is driven by a constant-current source I then the voltage across it can be expressed in decibels normalized to the peak value as

$$20 \log_{10} \left(\frac{|Z|I}{RI} \right). \tag{7.6}$$

This expression has a value of -3 dB when $|Z| = R/\sqrt{2}$ and then, from (7.5),

$$1 + Q^2 \left(\frac{\omega}{\omega_0} - \frac{\omega_0}{\omega} \right)^2 = 2. \tag{7.7}$$

The width of the curve is usually small so we can write

$$\omega = \omega_0 + \delta\omega,$$

where $\delta\omega \ll \omega_0$. Then, from (7.7),

$$\frac{\omega}{\omega_0} - \frac{\omega_0}{\omega} = \pm\frac{1}{Q} \tag{7.8}$$

so that

$$\left(1 + \frac{\delta\omega}{\omega_0}\right) - \frac{1}{\left(1 - \dfrac{\delta\omega}{\omega_0}\right)} = \pm\frac{1}{Q}. \tag{7.9}$$

Expanding the second bracket by the binomial theorem and neglecting powers of $\delta\omega/\omega_0$ higher than the first gives

$$\left(1 + \frac{\delta\omega}{\omega_0}\right) - \left(1 - \frac{\delta\omega}{\omega_0}\right) = \pm\frac{1}{Q} \tag{7.10}$$

or

$$\delta\omega \simeq \pm\frac{\omega_0}{2Q} \tag{7.11}$$

so that the width of the curve at the 3 dB points is ω_0/Q as shown in Fig. 7.2. Modern test equipment can display the response curve as shown with the vertical scale in decibels. This makes it easy to measure the Q factor of a resonator. Equation (7.11) provides a useful definition of the Q factor as

$$Q = \frac{\text{Resonant frequency}}{\text{Bandwidth}}. \tag{7.12}$$

Consideration of Fig. 7.1(a) and (7.4) shows that to obtain a high Q factor we must have a high shunt resistance. At radio frequencies resonant circuits made with lumped components generally have Q factors of the order of a few hundred. Microwave resonators commonly have Q factors of 1000 and can be made with values as high as 30 000 by careful design and manufacture. From these considerations it is evident that a high Q factor implies low losses and that leads to another useful way of defining Q.

When an alternating voltage $V = V_0 \exp j\omega t$ is applied to the terminals of the circuit shown in Fig. 7.1(a) the stored energy is

$$W_E = \tfrac{1}{2}CV_0^2. \tag{7.13}$$

It can be shown that this energy is transferred backwards and forwards between the capacitor and the inductor during each cycle with the total stored energy remaining constant provided that the voltage at the terminals is held constant. The mean rate of dissipation of energy is

$$P_L = V_0^2/2R. \tag{7.14}$$

Substituting these expressions into (7.4) gives

$$Q = \omega_0 W_E/P_L. \tag{7.15}$$

This equation can also be expressed as

$$Q = \frac{2\pi \times \text{Stored energy}}{\text{Energy dissipated per cycle}}. \tag{7.16}$$

The basic equivalent circuit of Fig. 7.1(a) is defined by the three parameters R, L and C. So far we have only generated two alternative parameters namely ω_0 and Q. The third is defined from (7.4) as

$$\left(\frac{R}{Q}\right) = \sqrt{\left(\frac{L}{C}\right)}. \tag{7.17}$$

Note that the value of (R/Q) is independent of the losses in the resonator. The physical significance of this parameter is revealed by substituting from (7.14) into (7.15) to give

$$\left(\frac{R}{Q}\right) = \frac{V_0^2}{2\omega_0 W_E}. \tag{7.18}$$

Thus (R/Q) is a measure of the relationship between the voltage across the resonator and the stored energy. In free-electron devices such as klystrons and particle accelerators the voltage across a resonator is used to interact with charged particles. A high (R/Q) means a high interaction voltage for a given stored energy so this parameter is a useful figure of merit.

To complete our review of the theory of parallel resonant circuits we must examine the phase of Z. From (7.2) we have

$$\angle Z = \arctan\left[Q\left(\frac{\omega_0}{\omega} - \frac{\omega}{\omega_0}\right)\right]. \tag{7.19}$$

This expression is zero at resonance so that Z is then wholly real as is clear from (7.2). As ω tends to zero the behaviour is dominated by the reactance of the inductor and $\angle Z$ tends to 90°. As ω tends to infinity the capacitor has the greater effect and $\angle Z$ tends to $-90°$. This behaviour is summarized in Fig. 7.3. The phase reversal at resonance is an important feature of the behaviour of any resonant device.

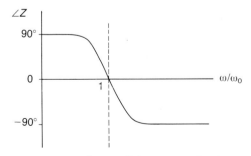

Fig. 7.3 Phase response curve of a parallel resonant circuit.

The properties of the series resonant circuit shown in Fig. 7.1b can be revealed by a similar analysis. The impedance of the network is

$$Z = R + j\omega L + \frac{1}{j\omega c} \tag{7.20}$$

which can be written

$$Z = R\left[1 + jQ\left(\frac{\omega}{\omega_0} - \frac{\omega_0}{\omega}\right)\right] \tag{7.21}$$

where

$$\omega_0 = 1/\sqrt{(LC)} \quad \text{and} \quad Q = \frac{1}{\omega_0 RC}. \tag{7.22}$$

This can be written, alternatively, as an admittance

$$Y = \frac{G}{1 + jQ\left(\frac{\omega}{\omega_0} - \frac{\omega_0}{\omega}\right)}, \tag{7.23}$$

where $G = 1/R$. Equation (7.23) is mathematically identical to (7.2) so all the conclusions for parallel resonant circuits can be applied to series resonant circuits if admittance is substituted for impedance. Hence a series resonant circuit has a maximum admittance G at resonance and a phase reversal exactly as shown in Figs. 7.2 and 7.3.

7.2 TRANSMISSION-LINE RESONATORS

In distributed circuits such as transmission lines and waveguides resonant behaviour is associated with the presence of standing waves. To illustrate this consider the three situations shown in Fig. 7.4. Figure 7.4(a) shows a length of transmission line with short circuits at either end. A voltage wave is reflected by a short circuit with 180° change of phase. If a wave starts from A with zero phase it has a phase of $-\theta$ on arrival at B. The reflected wave at B has a phase of $(\pi - \theta)$ which becomes $(\pi - 2\theta)$ on arrival at A. This wave after reflection has a phase of $(2\pi - 2\theta)$. For resonance to occur

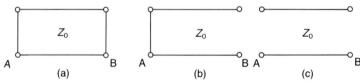

Fig. 7.4 Resonant sections of transmission line: (a) short circuit both ends, (b) short circuit and open circuit, and (c) open circuit both ends.

it is necessary for this wave to be in phase with the wave originally launched at A. The condition for resonance is therefore

$$(2\pi - 2\theta) = 0, -2\pi, -4\pi, \ldots, \text{etc.},$$

that is

$$\theta = n\pi, \tag{7.24}$$

where n is a positive integer. In other words the line is resonant when it is a whole number of half wavelengths long.

In the second case shown in Fig. 7.4(b) we recall that a voltage wave is reflected without change of phase by an open circuit. The phase of the wave which has been reflected at both B and A is therefore $(\pi - 2\theta)$. The condition for resonance is therefore

$$(\pi - 2\theta) = 0, -2\pi, -4\pi, \ldots, \text{etc.}$$
$$\theta = (n - \tfrac{1}{2})\pi, \tag{7.25}$$

so the line is an odd number of quarter wavelengths long at resonance.

In the third case the wave is reflected without a change of phase so that for resonance

$$2\theta = 2\pi, 4\pi, \ldots, \text{etc.}$$
$$\theta = n\pi \tag{7.26}$$

giving the same frequencies as in the first case.

The wave patterns on the lines are given by the sums of waves of equal amplitudes travelling in the positive and negative directions.

$$V = V_0 \exp j(\omega t - kx) \pm V_0 \exp j(\omega t + kx). \tag{7.27}$$

The sign of the amplitude of the wave travelling in the $-x$ direction is chosen so that the boundary condition at A is satisfied. Thus when there is an open circuit at A

$$V = 2V_0 \exp j\omega t \cos kx, \tag{7.28}$$

whilst with a short circuit at A

$$V = -2jV_0 \exp j\omega t \sin kx. \tag{7.29}$$

If the length of the line is L then the boundary conditions at B are satisfied if $\theta = kL$ is given by (7.24) when the two boundaries are the same and by (7.25) when they are different.

The voltage wave patterns at resonance are therefore

$$V = -2jV_0 e^{j\omega t} \sin\left(\frac{n\pi x}{L}\right) \tag{7.30}$$

$$V = -2\mathrm{j}V_0 \mathrm{e}^{\mathrm{j}\omega t} \sin\left(\frac{(n - \frac{1}{2})\pi x}{L}\right) \tag{7.31}$$

and
$$V = 2V_0 \mathrm{e}^{\mathrm{j}\omega t} \cos\left(\frac{n\pi x}{L}\right), \tag{7.32}$$

for the resonators shown in Fig. 7.4(a), (b) and (c), respectively, where $n = 1, 2, 3$, etc. Note that these are all standing waves, not travelling waves.

This discussion highlights one important difference between a lumped-element resonant circuit and a distributed resonant circuit. That is that a distributed resonant circuit has not one but an infinite set of resonances. Figure 7.5 shows the voltage wave patterns for the lowest three resonant modes of the lines in Fig. 7.4.

Example

Calculate the three lowest resonant frequencies for a 1 metre length of polythene insulated coaxial cable with both open- and short-circuit terminations.

Solution

The relative permittivity of polythene is 2.25 so the phase velocity is

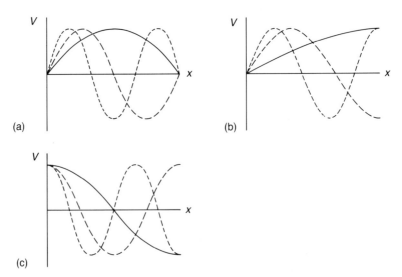

Fig. 7.5 Voltage standing waves for the lowest three resonances of each of the resonant transmission lines shown in Fig. 7.4.

$c/\sqrt{2.25} = 0.2 \times 10^9\,\mathrm{m\,s^{-1}}$. For an open circuit at one end and a short circuit at the other the resonant frequencies are given by

$$f = (n + \tfrac{1}{2})(v_{\mathrm{ph}}/2L) \tag{7.33}$$

so there are resonances at 50 MHz, 150 MHz and 250 MHz. Similarly when there are short or open circuit terminations at both ends the resonant frequencies are

$$f = n v_{\mathrm{ph}}/2L \tag{7.34}$$

giving resonances at 100 MHz, 200 MHz and 300 MHz.

This example makes the point that the lowest resonant frequencies of the lengths of cable commonly used in the laboratory are well down into the VHF and UHF regions. Thus if care is not taken with making correct terminations it is possible for resonances in circuits to be troublesome at quite low frequencies. It must be remembered that the sets of resonant frequencies are infinite so resonances can be detected in all higher-frequency bands.

The current waveform corresponding to (7.27) is given by

$$I = \frac{V_0}{Z_0}\,\mathrm{e}^{\mathrm{j}(\omega t - kx)} \mp \frac{V_0}{Z_0}\,\mathrm{e}^{\mathrm{j}(\omega t + kx)}, \tag{7.35}$$

where the reversal of the sign of the second term is necessary because the current in the wave travelling in the negative x direction must be opposite to that for the wave in the positive x direction. The current patterns corresponding to (7.30) to (7.32) are therefore

$$I = \frac{2V_0}{Z_0}\,\mathrm{e}^{\mathrm{j}\omega t}\cos\left(\frac{n\pi x}{L}\right) \tag{7.36}$$

$$I = \frac{2V_0}{Z_0}\,\mathrm{e}^{\mathrm{j}\omega t}\cos\left[\frac{(n - \tfrac{1}{2})\pi x}{L}\right] \tag{7.37}$$

$$I = \frac{-2\mathrm{j}V_0}{Z_0}\,\mathrm{e}^{\mathrm{j}\omega t}\sin\left(\frac{n\pi x}{L}\right). \tag{7.38}$$

Note that the currents and voltages are in phase quadrature in time implying that there is no net power flow in the resonator, also that they are in phase quadrature in space so that a voltage maximum corresponds to a current zero and vice versa. This must be clearly distinguished from the situation with a propagating wave where the current and voltage are in phase.

The voltage and current patterns are associated with distributions of the electric and magnetic fields around the line. Figure 7.6 shows how the field patterns change during one cycle. One way of thinking about this process is to imagine a static charge distribution set up as shown in Fig. 7.6(a). This

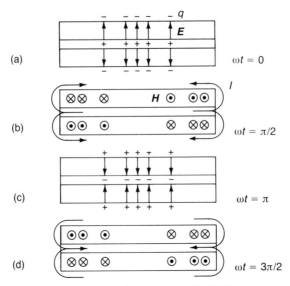

Fig. 7.6 Resonant section of coaxial line showing the fields, currents and charges at different times during the r.f. cycle.

distribution has an electric field associated with it as shown. The arrangement is unstable and so the positive and negative charges move towards each other producing currents and magnetic fields as shown in Fig. 7.6(b). Because the electrons possess inertia they overshoot producing a new charge distribution as in Fig. 7.6(c). This in turn is unstable and the cycle is continued.

So far we have assumed that the voltages and currents necessary to sustain the oscillation exist without saying how this might be achieved. The resonator shown in Fig. 7.6 is closed and ohmic damping would reduce the magnitude of the oscillation to zero over a few cycles. It follows that any resonator must be coupled to an external power source if sustained oscillation is to occur. The amplitude of oscillation excited is then just that for which the losses exactly match the incoming power.

Figure 7.7 shows a resonant circuit excited from a source of impedance Z_0 through a coupling capacitor. At resonance the impedance of the resonant circuit is just R and the output voltage is

$$V_{out} = \frac{RZ_0I_0}{Z_0 + R + jX} \tag{7.39}$$

using the current-splitting rule. The amplitude of the output voltage is greatest when $X = 0$ but then Z appears in parallel with R lowering the Q of the resonator. The loaded Q of the resonator is

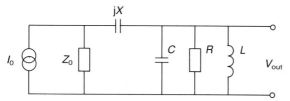

Fig. 7.7 Equivalent circuit of a parallel resonant circuit capacitively coupled to a source.

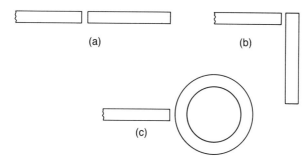

Fig. 7.8 Microstrip resonators: (a) end-coupled, (b) side-coupled, and (c) ring resonators.

$$Q_{\mathrm{L}} = \frac{1}{\omega_0 L}\left(\frac{RZ_0}{R + Z_0}\right), \tag{7.40}$$

where Z_0 is the impedance of the external loading of the resonator. Since one of the main uses of resonators is to provide frequency selectivity it is usually undesirable to allow the Q to be degraded in this way. Thus the coupling between the input line and the resonator is usually weak and X is correspondingly large.

The circuit of Fig. 7.7 is equivalent to the microstrip resonators shown in Fig. 7.8. The straight resonators shown in Figs. 7.8(a) and (b) must have lengths slightly less than half a wavelength to allow for the effects of fringing discussed in Section 6.2. The ring resonator must have a perimeter equal to one wavelength. The ring resonator has the advantage that it has a higher Q factor because it does not have the radiation losses which occur at the free ends of the straight resonators. A fuller discussion of microstrip resonators is given by Edwards (1981).

Coupling into the coaxial-line resonator shown in Fig. 7.6 can be achieved by means of an electric dipole at its mid-plane or a magnetic dipole at the end. These could be either wire or aperture dipoles. Figure 7.9 shows coupling arrangements with wire dipoles. The electric dipole (Fig. 7.9(a)) is placed at the plane of maximum electric field for optimum coupling. The equivalent circuit of this arrangement is as shown in Fig. 7.7. The magnetic

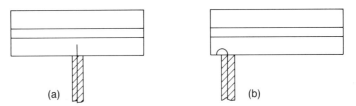

Fig. 7.9 Coupling into a coaxial-line resonator: (a) via the electric field, and (b) via the magnetic field.

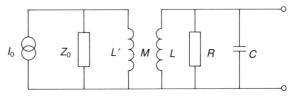

Fig. 7.10 Equivalent circuit for inductive coupling to a parallel resonant circuit.

dipole (Fig. 7.9(b)) is placed in a region of high magnetic field. The equivalent circuit of this arrangement is shown in Fig. 7.10. The strength of the coupling is adjusted by changing the size of the coupling loop or by turning so that it intercepts less of the flux circulating in the resonator. Either of these is equivalent to adjusting the mutual inductance M in Fig. 7.10. To couple selectively into a higher-order mode of the resonator the dipole is moved to a position which corresponds to a field maximum for that mode.

7.3 CAVITY RESONATORS

We have seen in Chapter 2 that metallic waveguides can be treated as transmission lines. It follows that resonators can be made out of short-circuited sections of waveguide. Open-circuit terminations are impossible to realize because an open end of waveguide radiates too well. For the moment let us consider only TE modes in the rectangular waveguide shown in Fig. 7.11. The electric fields for these modes are given by (2.60) and (2.61). The z variation is as $\exp \mathrm{j}(\omega t - k_g z)$, where

$$\left(\frac{n\pi}{a}\right)^2 + \left(\frac{m\pi}{b}\right)^2 + k_g^2 = k_0^2 \tag{7.41}$$

(equation (2.63)). When the ends of the waveguide are closed by metal walls separated by a distance c the boundary conditions in the z direction require that

$$k_g c = l\pi, \tag{7.42}$$

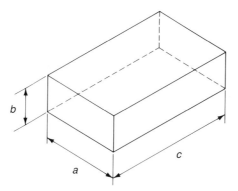

Fig. 7.11 Rectangular resonant cavity.

where $l = 1, 2, 3$, etc. Substituting this into (7.41) we find that the resonant frequencies are given by

$$\frac{\omega_0}{v_{ph}} = \left[\left(\frac{n\pi}{a}\right)^2 + \left(\frac{m\pi}{b}\right)^2 + \left(\frac{l\pi}{c}\right)^2\right]^{\frac{1}{2}}, \qquad (7.43)$$

where v_{ph} is the phase velocity of TEM waves in the medium filling the waveguide. When the field patterns of these modes are analysed it is found that the electric and magnetic fields are in phase quadrature in both time and space exactly as in the case of the coaxial-line resonator discussed above. The field patterns for a few of the lower resonant modes are shown in Fig. 7.12. The notation for these resonances is a bit tricky because it depends upon which direction is taken as the reference direction. For example the mode shown in Fig. 7.12(a) could be described as, for example, TE_{011} or TM_{110}. It is useful to acquire the ability to sketch the field patterns of different possible resonant modes of a cavity. It helps to think of the magnetic field as being generated by the displacement current associated with the electric field but it has to be remembered that they are in phase quadrature.

The fields of the lowest resonance of an air-filled cavity are given by

$$E_x = E_0 \sin\left(\frac{\pi y}{a}\right) \sin\left(\frac{\pi z}{c}\right) e^{j\omega t} \qquad (7.44)$$

$$H_y = \frac{j\pi E_0}{\omega\mu_0 c} \sin\left(\frac{\pi y}{a}\right) \cos\left(\frac{\pi z}{c}\right) e^{j\omega t} \qquad (7.45)$$

$$H_z = \frac{-j\pi E_0}{\omega\mu_0 a} \cos\left(\frac{\pi y}{a}\right) \sin\left(\frac{\pi z}{c}\right) e^{j\omega t} \qquad (7.46)$$

Integrating the square of the electric field gives the stored energy as

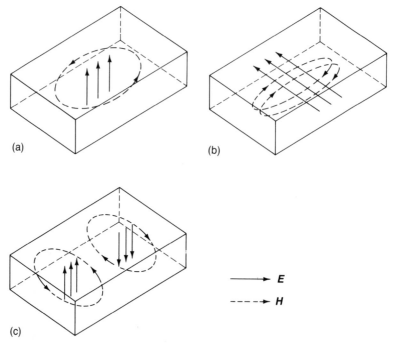

Fig. 7.12 Modes in a rectangular resonant cavity: (a) TE_{110}, (b) TE_{101} and (c) TE_{210}.

$$W_E = \frac{\varepsilon_0}{8} \, abc \, E_0^2, \qquad (7.47)$$

whence the (R/Q) of the mode is, from (7.18),

$$\left(\frac{R}{Q}\right) = \frac{4}{\pi} \sqrt{\left(\frac{\mu_0}{\varepsilon_0}\right)} \frac{b}{(a^2 + c^2)^{\frac{1}{2}}}. \qquad (7.48)$$

The previous discussion has concentrated on rectangular cavities because they are easiest to analyse mathematically. It must be remembered, however, that any closed conducting box will behave as a resonator with an infinite number of resonant modes. In general the lowest mode will have a wavelength which is of the same order of magnitude as the longest dimension of the box. Circular 'pill box' and re-entrant cavities as shown in Fig. 7.13 are in common use. An analytical solution for the pill-box cavity can be obtained in terms of Bessel functions (Appendix B). The lowest resonant frequency is given by

$$k_0 r = 2.405. \qquad (7.49)$$

The (R/Q) depends only on the cavity geometry. It can be computed from (7.6) using the exact solution for the electric field distribution. The result is

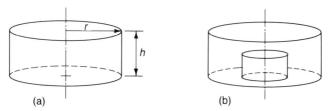

Fig. 7.13 Cylindrical resonant cavities: (a) pill-box, and (b) re-entrant cavities.

$$\left(\frac{R}{Q}\right) = 188.1\,\frac{h}{r}. \tag{7.50}$$

It is not possible to obtain exact analytical solutions for circular re-entrant cavities. Design curves are given by Saad (1971) and useful approximate formulae by Fujisawa (1958).

For more complicated cavity shapes it is necessary to make use of computer modelling techniques. Examples are the programs SUPERFISH (Halbach and Holsinger, 1976) and URMEL (Weiland, 1983) for cylindrically symmetrical cavities and MAFIA (Weiland, 1985) and TLM (Akhtarzad and Johns, 1975) for general three-dimensional cavity geometries.

In general, computations of cavity Q factors assume that the losses are small so that the current distribution in the walls is the same as that for a lossless cavity. The energy loss per cycle and the stored energy are computer by integrating the current and field distributions and Q is calculated from (7.15). At microwave frequencies the surface roughness is often comparable with the skin depth so the Q of a cavity depends upon surface finish as well as on the material from which the cavity is made. It is then necessary to use figures for surface resistance determined by experiment.

Coupling into cavities is achieved by using electric or magnetic dipoles arranged to couple to field maxima within the cavity exactly as in the case of the coaxial cavity discussed above. Figure 7.14 shows coupling from a waveguide to a cavity through an iris which acts as a magnetic dipole coupling element.

7.4 EFFECT OF RESONANCE ON SCREENED ENCLOSURES

We have already noted that electromagnetic power can be coupled into a screened enclosure through any small holes in the screen. The effects of such coupled power are made much worse if it happens to excite one of the resonances of the enclosure.

Example

Estimate the screening effectiveness of an aluminium box 1.0 mm thick whose dimensions are 50 mm × 100 mm × 200 mm at its lowest resonant

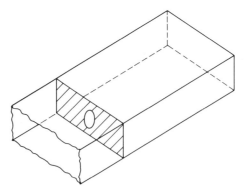

Fig. 7.14 Coupling from a waveguide to a resonant cavity via an iris.

frequency if there is a hole 5 mm in diameter in one side of the box. Assume that the Q of the resonance is 500.

Solution

The lowest resonant frequency of this box is given by (7.43) with $m = n = 1$ and $l = 0$. The result is 1.68 GHz. At this frequency the screening effectiveness of the enclosure without holes in it is very high.

The screening effectiveness of the enclosure can thus be calculated from the leakage through the hole using (6.16), (6.19) and (6.25) so that

$$A_a \simeq 27.3(1/5) = 5.5 \, \text{dB} \tag{7.51}$$

$$R_a \simeq -20 \log_{10} \left(\frac{2 \times 5}{179} \right) = 25.1 \, \text{dB} \tag{7.52}$$

and
$$B_a \simeq -20 \log_{10}(1 - 10^{-5.5}) = 2.9 \, \text{dB} \tag{7.53}$$

so that $S = 33.5 \, \text{dB}$. We recall that this is the ratio of the signal level inside the box to that which would exist if the box were not present. This figure assumes that the box is not resonant. If the power leaking into the box is P (assumed to be spread uniformly over the cross-sectional area of the box) then the mean electric field strength is

$$E_0 = \left[\frac{2PZ_0}{(0.05 \times 0.2)} \right]^{\frac{1}{2}}$$
$$= 275\sqrt{P}. \tag{7.54}$$

When the box is resonant the power is dissipated in the losses in the walls. For the dimensions given $(R/Q) = 26.8$ from (7.48) so that $R = 26.8 \times 500 = 13.4 \, \text{k}\Omega$. The voltage across the box is therefore

$$V_0 = E_r \times 0.05$$
$$= (2RP)^{\frac{1}{2}}, \tag{7.55}$$

whence the electric field strength at the centre of the resonant cavity is

$$E_R = 3274\sqrt{P}. \tag{7.56}$$

Comparing (7.54) and (7.56) shows that the effect of the resonance is to produce a substantial increase in the peak value of the electric field inside the box. The reduction in screening effectiveness is

$$S_R = 20 \log_{10} \left| \frac{E_R}{E_0} \right| = 21.5 \, \text{dB} \tag{7.57}$$

giving a final figure of 12 dB for the estimated screening effectiveness of the box at this frequency.

The size of the box in the preceding example was chosen to be typical of the kind of enclosure commonly used for electronic equipment. The lowest resonant frequency is high enough to cause little trouble unless the circuit enclosed is a microwave circuit. However if the box were larger, or if the circuit were encapsulated in epoxy resin the frequency would come down into the region of operation of high-speed digital circuits. In that case the box would screen the circuit very imperfectly from external signals at the resonant frequency. In addition, if the resonance were excited by the operation of the circuit itself at either the fundamental or a harmonic frequency, there could be strong cross-coupling effects within the circuit. This is why circuit designers must have an understanding of electromagnetic theory.

For much larger enclosures such as screened rooms there will be many higher-order resonances close to typical operating frequencies of electronic circuits. It is therefore necessary to ensure that these resonances are excited as weakly as possible by equipment within the room.

7.5 DIELECTRIC RESONATORS

In Chapter 3 it was demonstrated that strips and rods of dielectric material can act as waveguides. It follows that isolated pieces of dielectric may be expected to behave as resonators. One simple way of making a dielectric resonator is to place a section of dielectric rod waveguide between two parallel conducting planes. These act as short circuits and the lowest resonant frequencies can be worked out if the dispersion curve for the waveguide can be measured or calculated. If the relative permittivity of the material is high then (3.4) shows that signal is strongly reflected at an air–dielectric interface. For example titanium dioxide has a relative permittivity of 90 so the reflection coefficient of a TEM wave incident normally on the boundary from within the material is

$$\varrho = \frac{\sqrt{90} - 1}{\sqrt{90} + 1} = 0.81. \tag{7.58}$$

Thus it is possible to make a resonator out of such a material without any conducting boundaries. Note that (7.58) shows that the reflected wave is in phase with the incident wave so that, to a first approximation, the interface behaves as an open circuit. In order to achieve a reversal in the direction of the Poynting vector the direction of the magnetic field must be reversed at the boundary. That therefore behaves approximately as a magnetic short circuit or 'perfect magnetic conductor'.

If the approximation is made that the dielectric is bounded by a perfect magnetic conductor then the solution is obtained from that for a metallic resonant cavity by interchanging the roles of the electric and magnetic fields. In particular for a cylindrical piece of dielectric the lowest resonant mode has an axial magnetic field and a tangential electric field as shown in Fig. 7.15 by analogy with the pill-box cavity in Fig. 7.13. The frequency of this mode is then given by (7.49) with k_0 replaced by $k_1 = k_0\sqrt{\varepsilon_r}$, giving

$$f_{\text{GHz}} = \frac{0.115}{r\sqrt{\varepsilon_r}}. \tag{7.59}$$

Thus a titanium dioxide resonator is smaller than a cavity resonator having the same resonant frequency by a factor of $\sqrt{90} = 9.5$. Dielectric resonators are therefore much more compact than corresponding cavity resonators. They can be made with low loss and therefore high Q compared with microstrip resonators.

An important advantage of dielectric resonators is their temperature

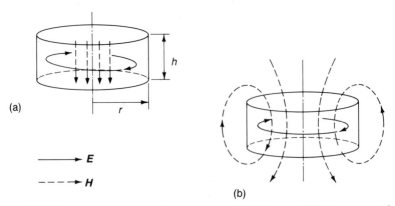

Fig. 7.15 Dielectric resonators: (a) approximate fields, and (b) the correct field pattern.

stability. The frequency of a metallic cavity resonator decreases with increasing temperature as the cavity expands. Dielectric resonators likewise expand as the temperature increases but the relative permittivity is also frequency dependent. For a material such as barium nanotitanate the relative permittivity decreases with temperature at a rate which almost exactly compensates for the thermal expansion. This is particularly useful for making stable local oscillators and narrow band filters.

The theory given above can only be approximate because, as (7.58) shows, the assumption that the boundaries are perfect magnetic conductors is not strictly correct. A full theory must include the effects of the leakage fields outside the dielectric. These are two-fold. First they modify the resonant frequency and, secondly, they result in a loss of energy by radiation so lowering the Q of the resonator. The field pattern around the resonator is, in practice, as shown in Fig. 7.15(b) so that it behaves as a magnetic dipole radiator. This is the mode commonly used in dielectric resonators. It is referred to as the $TE_{01\delta}$ or 'magnetic dipole' mode. An empirical formula for the resonant frequency obtained from numerical solutions is (from Kajfez and Guillon, 1986)

$$f_{GHz} = \frac{0.034}{r\sqrt{\varepsilon_r}} \left(\frac{r}{h} + 3.45 \right). \tag{7.60}$$

This formula is accurate to within 2% for $0.5 < r/h < 2$ and relative permittivities in the range 30 to 50.

The Q factor of an isolated dielectric resonator is about 50 because of radiation losses. If the resonator is placed within a conducting shield these losses are dramatically reduced and Q factors around 5000 can be achieved. Because in a typical case 95% of the stored electric energy and over 60% of the stored magnetic energy are within the dielectric the resonant frequency is determined largely by the dielectric so it is insensitive to small changes in the dimensions of the shield.

The leakage of magnetic flux from the resonator provides a convenient way of coupling it to a microstrip line. Figure 7.16 shows a typical arrangement.

It must be remembered that a dielectric resonator like any other distributed resonator has an infinite set of resonant modes. Sometimes these can be troublesome. For example Fig. 7.17 shows a simple form of waveguide window. Windows are used to enable microwave power to pass from air into vacuum in vacuum electron devices. Simple transmission-line theory shows that the window shown will be transparent at the frequency at which it is half a wavelength in thickness. The dielectric would usually be alumina which has a relative permittivity of 8.9. Whilst this figure is not as high as those for materials designed for use as dielectric resonators it is still high enough for dielectric resonances to occur. The shielded arrangement of the dielectric means that these modes can have quite high Q factors and be

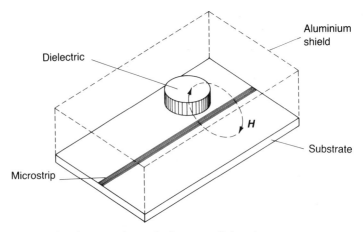

Fig. 7.16 Coupling from a microstrip line to a dielectric resonator.

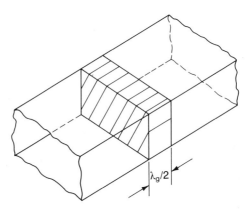

Fig. 7.17 Half-wavelength waveguide window.

only weakly coupled to the wave in the waveguide. They are quite difficult to detect and are therefore known as 'ghost modes' (Forrer and Jaynes, 1960). If one of these modes occurs within the operating frequency band of the window appreciable power can be dissipated by it, thus giving rise to unwanted heating and possible destruction of the window.

7.6 FABRY–PÉROT RESONATORS

A particularly simple form of resonator for TEM waves can be made by arranging a pair of plane reflecting surfaces parallel to each other as shown in Fig. 7.18. This arrangement is known as a Fabry–Pérot resonator. Res-

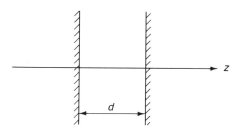

Fig. 7.18 Fabry–Pérot resonator.

onance occurs when the waves are travelling normal to the surfaces and the condition for resonance is

$$d = n\lambda/2. \tag{7.61}$$

If the surfaces are made so that the reflection is not quite perfect then it is possible to couple power into and out of the resonator. At microwave frequencies this could be achieved by using wire grids or thin conducting films evaporated on to dielectric surfaces. The former arrangement provides the possibility of making resonators which select one polarization of the wave. At optical frequencies the partial reflection is achieved by the use of thin films.

Fabry–Pérot resonators are used in lasers and are also important as a way of making accurate optical filters.

7.7 FILTER THEORY

Resonators are important because of their ability to select and reject frequencies. This property finds particular application in the fabrication of filters. A filter is a device which passes a band of frequencies whilst blocking other frequencies. Figure 7.19 shows the idealized transfer functions of the four possible types of filter. These are respectively, low-pass, high-pass, band-pass and band-stop filters. Their uses include the suppression of harmonics and the selection of bands of frequencies in frequency-domain multiplex communication systems. The theory of filters including techniques for their synthesis is a major subject in its own right. Here we shall concentrate on the links between filter theory and the properties of transmission lines and resonators already discussed.

At low frequencies filters can be realized using lumped inductors and capacitors. Figure 7.20 shows examples of lumped-element low- and high-pass filter networks. The way in which these work can easily be understood if it is remembered that an inductor has an impedance which is zero for d.c. and which increases with frequency. Similarly, a capacitor provides a total block to d.c. but has a decreasing impedance as the frequency increases.

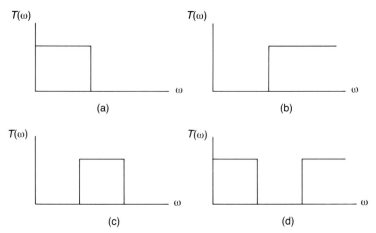

Fig. 7.19 The four ideal filter characteristics: (a) low-pass, (b) high-pass, (c) band-pass, and (d) band-stop filters.

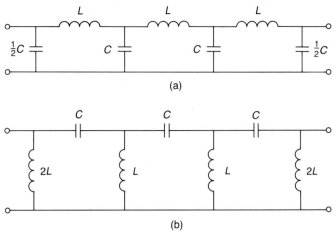

Fig. 7.20 Filter networks: (a) low pass, and (b) high pass.

Thus the low-pass filter shown in Fig. 7.20(a) provides no obstacle to d.c. At high frequencies the passage of current is blocked by the inductors and a low impedance path to ground is provided by the capacitors. It can be shown (Jones and Hale, 1982) that the transfer function of the network shown in Fig. 7.20(a) can be designed to have the forms

$$|T(\omega)|^2 = \frac{1}{1 + k^2\left(\dfrac{\omega}{\omega_0}\right)^{2N}} \qquad (7.62)$$

for a Butterworth response and

$$|T(\omega)|^2 = \frac{1}{1 + k^2 T_N^2\left(\dfrac{\omega}{\omega_0}\right)} \qquad (7.63)$$

for a Tchebychev response by a suitable choice of the component values. In these expressions k is a parameter which defines the minimum loss within the passband of the filter, N is the number of components in the network, T_N is the Tchebychev polynomial of order N and ω_0 is given by

$$\omega_0 = 1/\sqrt{(LC)}. \qquad (7.64)$$

It is also necessary to ensure that the component values are chosen so that the filter is matched to the source impedance within its pass band.

A high-pass filter can be derived from a low-pass filter by the transformation

$$\frac{\omega}{\omega_0} \rightarrow -\frac{\omega_0}{\omega}. \qquad (7.65)$$

Thus in Fig. 7.20 the series inductance L in Fig. 7.20(a) transforms to

$$-j\,\frac{\omega_0^2 L}{\omega} = \frac{1}{j\omega C}, \qquad (7.66)$$

where C is the corresponding shunt capacitance in Fig. 7.20(b).

Figure 7.21 shows typical transfer characteristics for a pair of low- and high-pass filters designed in this way.

Band-pass and band-stop filters can also be derived from a low-pass filter by the transformations

$$\frac{\omega}{\omega_0} \rightarrow \frac{\omega_0}{\omega_2 - \omega_1}\left(\frac{\omega}{\omega_0} - \frac{\omega_0}{\omega}\right) \qquad (7.67)$$

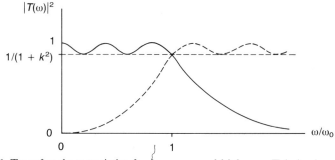

Fig. 7.21 Transfer characteristics for low-pass and high-pass Tchebychev filters.

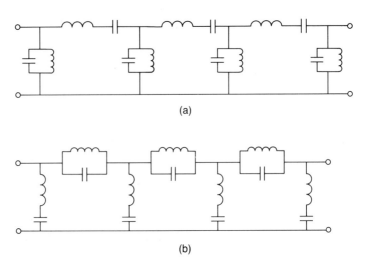

Fig. 7.22 Filter networks: (a) band pass, and (b) band stop.

and

$$\frac{\omega}{\omega_0} \rightarrow \left[\frac{\omega_0}{\omega_2 - \omega_1} \left(\frac{\omega}{\omega_0} - \frac{\omega_0}{\omega} \right) \right]^{-1}, \qquad (7.68)$$

where ω_1 and ω_2 are the upper and lower limits of the pass or stop band and $\omega_0 = \sqrt{(\omega_1\omega_2)}$. The second term in each of these expressions is familiar from equations (7.2) and (7.21) so it is not unexpected that these two filters incorporate resonant elements. It can be shown that the transformations replace the series inductors of the low-pass filter by a series resonant circuit and the capacitors by parallel resonant circuits to obtain the band-pass network shown in Fig. 7.22(a). Exchanging the series and shunt resonant circuits produces the band-stop network shown in Fig. 7.22(b). Once again it is possible to understand how these networks work if it is recalled that the shunt and series resonant circuits have maximum and minimum impedance respectively at resonance.

7.8 TRANSMISSION-LINE FILTERS

Transmission-line filters are usually realized in microstrip or suspended substrate stripline. For convenience the discussion here will assume that microstrip is being used.

The basic elements from which stripline filters are made up are the short- and open-circuit stubs shown in Fig. 7.23. It is easily shown that the input impedances are

$$Z = jZ_0 \tan \theta \qquad (7.69)$$

for the short-circuit stub, where θ is the electrical length, and

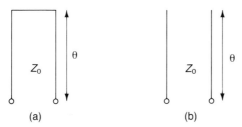

Fig. 7.23 Transmission-line filter elements: (a) short-circuit stub, and (b) open-circuit stub.

$$Z = \frac{1}{j Y_0 \tan \theta} \tag{7.70}$$

for the open-circuit stub. Thus the inductors and capacitors of the low-frequency networks can be replaced by short- and open-circuit stubs, respectively. Tan θ, which is frequency dependent, takes the place of ω. This produces one important difference between microwave filters and lumped-element filters. Tan θ is periodic in ω so there is a succession of frequencies at which a stub will behave as a short circuit or an open circuit. At intermediate frequencies the roles of the stubs are reversed. The result of this is that a low-pass network has a succession of higher-order pass bands with stop bands between them. Figure 7.24 shows the stripline analogue of the low-pass network shown in Fig. 7.20(a). The transfer factor of this filter is shown in Fig. 7.25. The filter shown in Fig. 7.24 is impossible to realize because of the difficulty of making the shunt stubs and because of the need to have a finite distance between the points at which the stubs are connected together. The solution to this problem is to make use of the impedance transformation properties of the main transmission line to allow all the elements to be open-circuit shunt stubs.

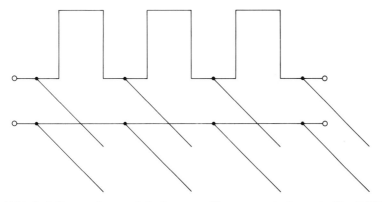

Fig. 7.24 Stripline analogue of the low-pass filter network shown in Fig. 7.20(a).

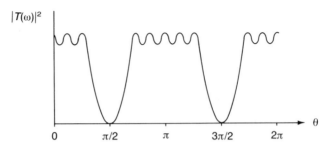

Fig. 7.25 Periodic frequency response of a Tchebychev low-pass filter realized in stripline.

The theory of a simple filter comprizing a series of reflecting elements separated by equal electrical lengths θ can be understood by reference to the discussion of broadband matching in Section 6.7. There it was shown that, for small reflections, the reflection coefficient of the set of reflections $\varrho_1, \varrho_2, \varrho_2, \varrho_1$ could be written

$$|\varrho| = |8\varrho_1 x^3 + 2(\varrho_2 - 3\varrho_1)x| \qquad (7.71)$$

(equation (6.3)) where $x = \cos\theta$. It was also shown that this could be put into the form of Butterworth or Tchebychev polynomials by a suitable choice of ϱ_1 and ϱ_2. Now the power reflected by this device is given by

$$P_r = |\varrho|^2 P_i \qquad (7.72)$$

so the power transmitted is given by

$$\frac{P_t}{P_i} = 1 - |\varrho|^2$$

$$\simeq \frac{1}{1 + |\varrho|^2} \qquad (7.73)$$

which can be expressed in the forms given in (7.62) and (7.63). A very similar result could be obtained by using a stepped impedance transformer as shown in Fig. 7.26(a). Another possible filter configuration is the parallel coupled band-pass filter shown in Fig. 7.26(b). Each of the sections of line behaves as a parallel resonant circuit since there is a voltage maximum at its ends when it is resonant. The coupling between the elements is capacitive so the device has the equivalent circuit shown in Fig. 7.26(c). Comparison between this circuit and that in Fig. 7.22(a) shows that the filter must be a band-pass filter. All these filters are examples of filters with periodic structures.

The design of microstrip filters is a major subject in its own right and the reader should refer to specialized texts (Matthaei *et al.*, 1964) for further information.

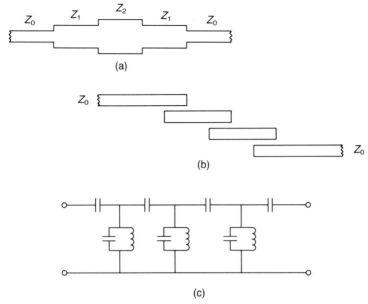

Fig. 7.26 Microstrip filters: (a) stepped-impedance, and (b) parallel-coupled band-pass filters with (c) the equivalent circuit of the latter.

7.9 WAVEGUIDE FILTERS

At power levels above a few watts it is necessary to use waveguide technology to make filters. The approach used is essentially the same as in the case of transmission lines.

Figure 7.27(a) shows a section of waveguide with a resonant section formed by two transverse walls at A and B. These walls have small coupling irises in them. An incident wave in the waveguide is almost completely reflected at A producing a standing wave in the waveguide. The tangential magnetic field therefore induces a magnetic dipole in the iris at A. The field of this dipole at the iris at B is negligible unless the cavity is resonant. Under resonant conditions the field of the cavity excites the iris at B which then radiates power into the output waveguide. This arrangement therefore acts as a band-pass filter. Provided that the coupling irises are small the properties of the cavity are not affected by loading by the input and output waveguides and the transfer characteristic of the filter is essentially that of Fig. 7.2.

The problem with this kind of filter is that the transmission loss is rather large because of the bad mismatch at A. If the irises are enlarged to increase the coupling then the input and output guides load the cavity reducing its Q factor and possibly tuning its resonant frequency.

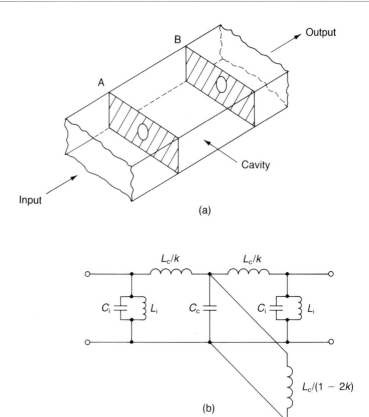

Fig. 7.27 Iris-coupled waveguide band-pass filter: (a) general arrangement, and (b) equivalent circuit.

The equivalent circuit of the filter is shown in Fig. 7.27(b). The irises are represented by parallel LC circuits because there is inductance associated with the flow of current around the hole and capacitance associated with the displacement current across it. The cavity inductance L_c is divided into three parts because not all the circulating current in the cavity is intercepted by the irises. The resulting equivalent circuit clearly has a band-pass characteristic as can be seen by comparing it with Fig. 7.22(a).

In order to control the transfer characteristic of the filter several cavities may be connected in series as shown in Fig. 7.28. The behaviour of this arrangement can be understood by considering the extreme cases when the cavities are resonating in phase or in antiphase with each other. Figure 7.29 shows how this happens. In Fig. 7.29(a) the cavities are excited in phase with each other. The conduction currents in adjacent cavities are crossing the irises in the same directions. The irises being small are well below their lowest resonant frequency so they present an inductance to the current.

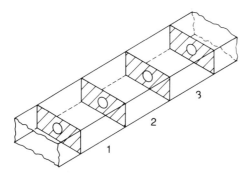

Fig. 7.28 Multi-cavity waveguide band-pass filter.

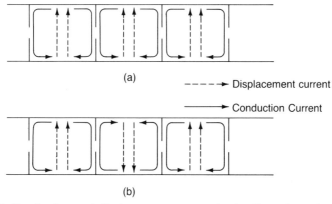

Fig. 7.29 Conduction and displacement currents in the filter shown in Fig. 7.28 (a) at zero phase shift per section, and (b) at 180° phase shift per section.

This inductance is in series with the cavity inductance so it has the effect of lowering the resonant frequency a little. When the cavities are excited in antiphase (Fig. 7.29(b)) the conduction currents flow across the irises in opposite directions. There is, therefore, no net current flow across each iris and the resonant frequency is unperturbed. For other possible phase changes there is an intermediate situation so the in-phase and anti-phase conditions represent the edges of a band of frequencies which the filter will pass. Careful design of the sizes of the irises and of the resonant frequencies of the cavities can produce either Butterworth or Tchebychev characteristics in the pass band.

The filter shown in Fig. 7.28 also has a series of higher-order pass bands corresponding to those higher-order modes of the cavities which couple to the irises. Thus a filter comprising a series of weakly coupled cavities has narrow pass bands separated by broad stop bands. In describing the action

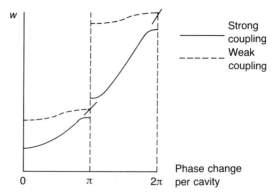

Fig. 7.30 The dispersion diagram of the filter shown in Fig. 7.28.

of the filter in this way we have approached the general problem of the coupled resonator filter from the point of view of weak coupling. The same qualitative results were obtained in the previous section by considering the case of a transmission line with small regularly spaced perturbing objects.

This kind of behaviour is found whenever waves propagate through periodic structures. Figure 7.30 (in which frequency is plotted against phase change per cavity) shows an alternative way of displaying the same information. For weak coupling between the resonators the pass bands are narrow and are very close to the resonant frequencies of the cavities. For strong coupling the dispersion curve is very similar to that of an empty waveguide expcept where the separation between the discontinuities is an integral number of half wavelengths. Then it is discontinuous so that there is a stop band.

This discontinuity can be explained by considering Fig. 7.31 which shows the electric and magnetic fields in the waveguide when the separation of the obstacles is a half wavelength. The cumulative reflections set up a standing wave in the guide and any general standing wave can be regarded as the superposition of the two 'normal modes' shown in the figure. In Fig.

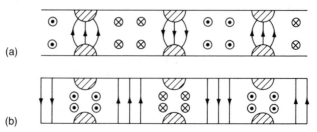

Fig. 7.31 Fields in the filter shown in Fig. 7.28 at the two edges of the stop band when the phase change per section is 180°.

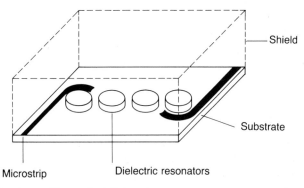

Fig. 7.32 Band-pass filter using dielectric resonators.

7.31(a) the electric field maxima coincide with the obstacles. For this mode the capacitance is reduced and the frequency is slightly higher than in the unperturbed mode. In the second case (Fig. 7.31(b)) the magnetic field is perturbed producing an increase in the effective inductance and a slight reduction in the resonant frequency.

We have therefore demonstrated, both from the point of view of coupled resonators and from that of perturbed waveguide modes, that periodic arrays of discontinuities in a waveguide produce alternate pass and stop bands whose widths depend upon the magnitudes of the discontinuities.

Stripline filters have the disadvantage that it is difficult to get very high values of Q. The theoretical behaviour is therefore limited by the inherent bandwidths of the elements from which the filter is made up. One solution to this is to make use of dielectric resonators as shown in Fig. 7.32. The resonators are inductively coupled just like the cavities in Fig. 7.28. The filter has a band-pass characteristic.

7.10 OPTICAL FILTERS

At optical frequencies filters can be made by using Fabry–Pérot resonators. A series of layers of transparent dielectric separated by thin metallic films or by layers of a different dielectric has the equivalent circuit shown in

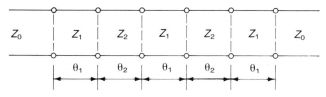

Fig. 7.33 Equivalent circuit for an optical filter employing layers of different dielectric materials.

Fig. 7.33. This periodic structure has alternate pass and stop bands exactly like those discussed in the preceding sections. By careful choice of the filter dimensions it is possible to ensure that all the pass bands except one lie outside the visible spectrum. This filter is a band-pass filter. Further details can be found in books on optics (Longhurst, 1973).

7.11 CONCLUSION

In this chapter we have considered how resonant structures are made in transmission lines and waveguides. It has been demonstrated that such structures can be described in terms of lumped-element equivalent circuits with the difference that distributed structures can support infinite sets of higher-order resonant modes. Resonant circuits are important because of their frequency-selective properties. These can be enhanced by combining resonant circuits into filter structures. The four basic filter types (low pass, high pass, band pass and band stop) can all be derived from a low-pass filter. Filters can be designed in both transmission lines and waveguides from lumped-component, low-frequency, models. Certain changes are necessary in order to adapt the designs to the particular technology in which they are to be realized.

Most filters have a periodic structure and exhibit alternate pass and stop bands. The pass bands get wider as the strength of the coupling between the elements of the filter increases. This behaviour can be understood either by considering the perturbation of waves by regularly spaced discontinuities or by considering the behaviour of weakly coupled resonators.

EXERCISES

7.1 Calculate the equivalent circuit parameters for a microwave cavity which resonates at 3.40 GHz with a Q of 600 and an R/Q of 450 Ω. Find the amplitude and phase of its impedance at 3.38 GHz and 3.43 GHz.

7.2 Calculate the lowest three resonant frequencies of a 75 Ω semi-airspaced, coaxial cable ($\varepsilon_r = 1.06$) used as the downlead from a television antenna if the cable is 15 m long.

7.3 Calculate the loaded Q and the bandwidth of the resonator of Question 7.1 when it is connected to a 10 kΩ load.

7.4 Calculate the lowest three resonant frequencies of a rectangular cavity 17 mm $\times$ 21 mm $\times$ 12 mm. What are the new frequencies if the box is completely filled with epoxy resin ($\varepsilon_r = 3.5$)?

7.5 A rectangular metal box 510 mm $\times$ 550 mm $\times$ 126 mm which has a screening effectiveness of 120 dB is illuminated uniformly by an electromagnetic wave having a power density of $1 \, \text{mW m}^{-1}$ at the frequency of the lowest resonance of the box. If the Q factor of the resonance is 90 estimate the screening effectiveness at this frequency.

7.6 Calculate the dimensions of a pill-box dielectric resonator made of barium titanate ($\varepsilon_r = 1200$) if the resonant frequency is 9.5 GHz and the radius is twice the height.

Ferrite devices

8.1 INTRODUCTION

In Chapter 1 we saw that certain media known as gyromagnetic media have the property that positive and negative circular polarized waves have different propagation constants. These propagation constants can be altered by changing the static magnetic field applied to the medium. In this chapter we examine the ways in which the gyromagnetic properties of ferrites are used to make non-reciprocal devices and devices whose properties can be varied electrically.

8.2 MICROWAVE PROPERTIES OF FERRITES

Electrons possess a magnetic dipole moment but, in most materials, the electrons are paired in such a way that their dipole moments are opposite and the net dipole moment is zero. The special properties of ferrites arise because they contain unpaired electrons which are, therefore, free to respond to external magnetic fields.

Quantum theory (Bleaney and Bleaney, 1976) shows that an electron has a magnetic moment $m = 9.27 \times 10^{-24}\,\text{A m}^2$ and an angular momentum $P = 0.527 \times 10^{-34}\,\text{J s}$; these vectors are parallel to each other but in opposite directions. The ratio of the magnetic moment to the angular momentum is called the gyromagnetic ratio. It can be written

$$\gamma = \frac{m}{P} = g\,8.79 \times 10^{10}\,\text{c kg}^{-1}. \tag{8.1}$$

The constant g is known as the Lande factor. It has a value of 2 for free electrons and between 1.9 and 2.1 for electrons in most ferrites.

When an electron is placed in a static magnetic field the torque exerted on it by the field interacts with the angular momentum as shown in Fig. 8.1. The torque is

$$\boldsymbol{T} = \boldsymbol{m} \wedge \boldsymbol{B}_0 = -\gamma \boldsymbol{P} \wedge \boldsymbol{B}_0 \tag{8.2}$$

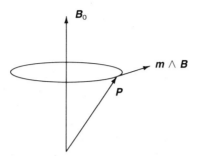

Fig. 8.1 Precession of the magnetic dipole moment of an electron around the direction of a static magnetic field.

and the equation of motion is

$$\frac{\mathrm{d}\boldsymbol{P}}{\mathrm{d}t} = -\gamma \boldsymbol{P} \wedge \boldsymbol{B}_0. \tag{8.3}$$

The rate of change of angular momentum is constant and directed at right angles to both $\boldsymbol{B}_0$ and $\boldsymbol{P}$. The tip of the vector $\boldsymbol{P}$ therefore traces a circle about the direction of the magnetic field with an angular velocity, known as the Larmor frequency,

$$\omega_0 = \gamma B_0 \tag{8.4}$$

which is independent of the angle between $\boldsymbol{P}$ and $\boldsymbol{B}_0$. The behaviour of the electrons is very like that seen when the axis of a spinning top moves in circles around the vertical. On a macroscopic scale the contributions of the individual dipoles can be summed to give a total dipole moment $\boldsymbol{M}$ per unit volume. The magnetic flux density within the material can then be written

$$\boldsymbol{B} = \mu_0(\boldsymbol{H}_0 + \boldsymbol{M}), \tag{8.5}$$

where $\boldsymbol{H}_0$ is the magnetizing field. Making use of (8.1) we can write, by analogy with (8.3)

$$\frac{\mathrm{d}\boldsymbol{M}}{\mathrm{d}t} = -\gamma\mu_0\boldsymbol{M} \wedge \boldsymbol{H}. \tag{8.6}$$

Now suppose that an small a.c. field is superimposed upon the static field so that

$$\boldsymbol{H} = \boldsymbol{H}_0 + \boldsymbol{H}_1 e^{j\omega t} \tag{8.7}$$

and

$$\boldsymbol{M} = \boldsymbol{M}_0 + \boldsymbol{M}_1 e^{j\omega t}. \tag{8.8}$$

Substituting these expressions into (8.6) we obtain

$$\frac{dM_0}{dt} + j\omega M_1 e^{j\omega t} = -\gamma\mu_0(M_0 \wedge H_0) - \gamma\mu_0(M_0 \wedge H_1)e^{j\omega t}$$
$$- \gamma\mu_0(M_1 \wedge H_0)e^{j\omega t}, \tag{8.9}$$

where second-order terms have been neglected. Equating the terms with $e^{j\omega t}$ dependence on the two sides of the equation gives

$$j\omega M_1 = -\gamma\mu_0(M_0 \wedge H_1) - \gamma\mu_0(M_1 \wedge H_0). \tag{8.10}$$

If the static field is aligned with the z axis and the field strength is large enough to magnetize the material to saturation then M_0 must also be parallel to the z axis. Equation (8.10) can then be written in component form as

$$j\omega M_x + \omega_0 M_y = \omega_M H_y \tag{8.11}$$

$$-\omega_0 M_x + j\omega M_y = -\omega_M H_x \tag{8.12}$$

$$j\omega M_z = 0, \tag{8.13}$$

where $M_1 = (M_x, M_y, M_z)$, $H_1 = (H_x, H_y, H_z)$ and $\omega_M = \mu_0\gamma M_0$.

Rearranging equations (8.11) and (8.12) gives

$$M_x = \chi H_x + j\varkappa H_y \tag{8.14}$$

$$M_y = -j\varkappa H_x + \chi H_y, \tag{8.15}$$

where

$$\chi = \frac{\omega_0\omega_M}{\omega_0^2 - \omega^2} \tag{8.16}$$

and

$$\varkappa = \frac{\omega\omega_M}{\omega_0^2 - \omega^2}. \tag{8.17}$$

Substituting (8.14) and (8.15) into (8.5) yields the matrix equation

$$\begin{pmatrix} B_x \\ B_y \\ B_z \end{pmatrix} = \mu_0 \begin{pmatrix} 1 + \chi & j\varkappa & 0 \\ -j\varkappa & 1 + \chi & 0 \\ 0 & 0 & 1 \end{pmatrix} \begin{pmatrix} H_x \\ H_y \\ H_z \end{pmatrix}. \tag{8.18}$$

This equation shows that in a magnetized ferrite material the directions of the small-signal B and H vectors are not the same as each other. The matrix (8.18) is called the tensor permeability of the material.

The preceding theory has assumed that there are no losses. In fact ferrites are lossy at microwave frequencies. It has been found that the effects of loss can be represented by replacing ω_0 in (8.16) and (8.17) by $(\omega_0 + j\omega\alpha)$, where α is a constant.

Studies of the propagation of plane TEM waves through a magnetized ferrite show that circularly polarized waves propagating in the direction of the static magnetic field have interesting properties. Consider, therefore, a circularly polarized wave whose magnetic field vector is given by

$$H_\pm = H_c(\hat{x} \mp j\hat{j}) \tag{8.19}$$

(see Section 1.7). We will assume that the magnetic field vector is parallel to the x, y plane. When (8.19) is substituted into (8.18) the result is

$$B_x = \mu_0(1 + \chi \pm \varkappa)H_c$$
$$B_y = \pm j\mu_0(1 + \chi \pm \varkappa)H_c \tag{8.20}$$

so that

$$B_\pm = \mu_\pm H_\pm, \tag{8.21}$$

where

$$\mu_\pm = \mu_0(1 + \chi \pm \varkappa). \tag{8.22}$$

The importance of this result is that for circularly polarized waves the permeability is a scalar not a tensor. Moreover its value is different for positive and negative polarizations. From (8.16), (8.17) and (8.22) we obtain

$$\mu_\pm = \mu_0\left(1 + \frac{\omega_M}{\omega_0 \mp \omega}\right). \tag{8.23}$$

This equation reveals a very important property namely that there is a resonance in the permeability for positive polarized waves but not for negative polarization. When the effects of losses are included (8.23) becomes

$$\mu_\pm = \mu_0\left[1 + \frac{(\omega_M/\omega_0)}{1 \mp \dfrac{\omega}{\omega_0}(1 \mp j\alpha)}\right]. \tag{8.24}$$

This can be written in terms of real and imaginary parts as

$$\mu_\pm = \mu'_\pm + j\mu''_\pm. \tag{8.25}$$

Figure 8.2 shows how these quantities vary with normalized frequency for typical values of ω_0 and ω_M. The propagation constants of circularly polarized waves are given by

$$k_\pm = \omega\sqrt{(\varepsilon\mu_\pm)}. \tag{8.26}$$

From this equation and Fig. 8.2 it can be seen that at most frequencies the propagation constants of the two waves are markedly different. Moreover the negatively polarized wave is virtually lossless at all frequencies whilst the positively polarized wave suffers strong attenuation at frequencies close to resonance. These properties are exploited in a variety of microwave devices described in the remainder of this chapter.

The theory of ferrites given in this section is only strictly applicable to a homogeneous material. In practice, ferrites are normally made as sintered blocks of polycrystalline material. The number of magnetic domains con-

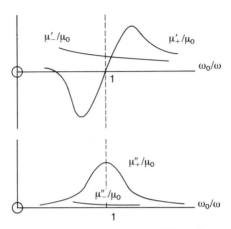

Fig. 8.2 Real and imaginary parts of the permeability of a magnetized ferrite for positive and negative circularly polarized waves.

tained in each crystal is determined by its size. The movement of domain walls within the crystals introduces losses and, for a material magnetized to saturation, this broadens the resonance somewhat but does not increase the propagation losses very much. When the material is not magnetized to saturation the effects of the different directions of magnetization of the domains must be taken into account by statistical methods. For further information on the properties of ferrites see Baden-Fuller (1987).

8.3 RESONANCE ISOLATORS

A very useful microwave device known as an isolator exploits the difference in the attenuation of positive and negative circularly polarized waves. The magnetic field components in a rectangular waveguide are given by

$$H_y = \frac{E_0}{Z_0} \frac{\lambda_0}{\lambda_g} \sin\left(\frac{\pi y}{a}\right) \cos\left(\omega t - k_g z\right) \tag{8.27}$$

$$H_z = \frac{E_0}{Z_0} \frac{\lambda_0}{\lambda_c} \cos\left(\frac{\pi y}{a}\right) \sin\left(\omega t - k_g z\right) \tag{8.28}$$

from (2.46) and (2.47). These fields are in phase quadrature so they constitute a circularly polarized field at a plane where their amplitudes are equal, that is

$$\frac{1}{\lambda_g} \sin\left(\frac{\pi y}{a}\right) = \pm \frac{1}{\lambda_c} \cos\left(\frac{\pi y}{a}\right) \tag{8.29}$$

or $$\tan\left(\frac{\pi y}{a}\right) = \pm \frac{\lambda_g}{2a}. \tag{8.30}$$

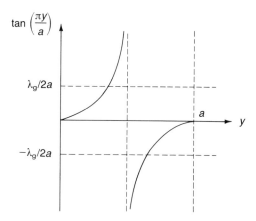

Fig. 8.3 Graphical solution of equation (8.30).

The solutions to this equation are indicated graphically in Fig. 8.3. There are two symmetrically placed planes parallel to the narrow wall of the waveguide for which (8.29) is satisfied. Now the direction of the circular polarization depends upon the direction of power flow in the waveguide. So, if a ferrite sheet is inserted in the waveguide at one of the planes indicated in Fig. 8.3 and subjected to a static magnetic field in the x direction, then waves travelling in one direction will be attenuated much more strongly than those travelling in the other direction. The magnitude of the static magnetic field must be chosen so that the ferrite is resonant at the signal frequency. Figure 8.4(a) shows the arrangement of a ferrite resonance isolator of this kind.

The isolator shown in Fig. 8.4(a) is unsatisfactory in a number of ways. The plane at which the ferrite must be placed depends on frequency as can be seen from Fig. 8.3 and the resonance band of the material is narrow. The device illustrated is, therefore, inherently narrow band. The presence of the ferrite perturbs the fields so that they are not exactly like those in an empty waveguide. Finally the thin sheet of ferrite is unable to dissipate much heat and the properties of the material are dependent on temperature. A better arrangement is that shown in Fig. 8.4(b) where the ferrite is in the form of strips in good thermal contact with the wall. The magnetic field is arranged so that it varies over the ferrite by the use of air gaps. In this way different parts of the ferrite are resonant at different frequencies and the device can be made broad band. Dielectric material is also sometimes used to ensure that the fields are truly circularly polarized. Isolators which work on this principle can be obtained which cover the entire frequency band of the waveguide with forward and reverse insertion losses of 1 dB and 30 dB respectively.

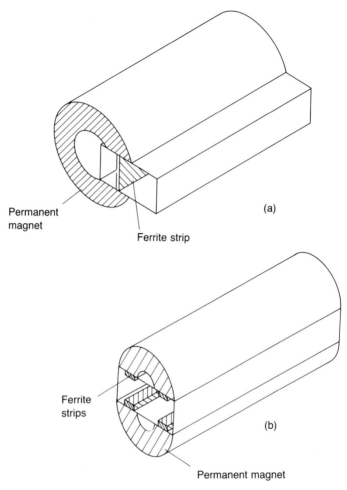

Permanent
magnet

(a)

Ferrite strip

Ferrite
strips

(b)

Permanent magnet

Fig. 8.4 Two types of ferrite resonance isolator.

8.4 PHASE SHIFTERS AND CIRCULATORS

Away from resonance the dependence of the permeability of a ferrite
material with magnetic field can be used to make a variable phase shifter.
In waveguide the arrangement is similar to that shown in Fig. 8.4(b) with
the permanent magnet replaced by an electro-magnet. The phase shift is
different for waves travelling in opposite directions because of the difference between the effective permeability for positively and negatively
polarized waves. When such a device is designed to produce a difference of
180° between the phase shifts for forward and backward waves it is known
as a gyrator.

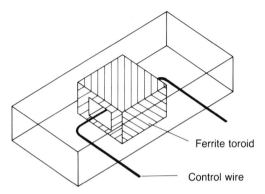

Fig. 8.5 Arrangement of a ferrite latching phase shifter.

If a fixed, switchable, phase change is required a latching phase shifter may be used. This device is shown in Fig. 8.5. A ferrite toroid is placed within a rectangular waveguide with its vertical sections arranged in the region of circularly polarized field. A control wire passes through the toroid enabling it to be magnetized in either direction by a pulse of current. The phase shift produced is adjusted by changing the length of the ferrite toroid.

A circulator is a device with three or more ports which has the property that a signal injected at one port emerges from the next one. This is shown diagrammatically in Fig. 8.6(a). One important application of this device is

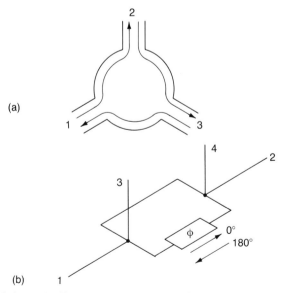

Fig. 8.6 (a) Schematic diagram of a circulator, (b) arrangement of a phase-shift circulator.

in separating the transmitted and received signals in a microwave communication system. In that case port 1 would be connected to the transmitter, port 2 to the antenna and port 3 to the receiver. Ideally the circulator would ensure that none of the transmitter power would find its way back to the receiver. A number of types of circulator exist which use the properties of ferrites in different ways.

Figure 8.6(b) shows a phase-shift circulator. A signal which enters at port 1 is split by a hybrid tee. The signal in one arm passes through a gyrator which has a phase shift of zero in the forward direction. The signals in the two arms meet at a second hybrid tee and, because they are in phase with each other, the combined signal emerges at port 2. If the signal is injected at port 2 the gyrator causes the signals in the two arms to be in antiphase so that the recombined signal emerges at port 3. Signals injected at ports 3 and 4 produce signals in antiphase in the two arms which recombine to give outputs at ports 4 and 1 respectively. The device therefore behaves as a four port circulator.

8.5 JUNCTION CIRCULATORS

A particularly compact form of circulator can be made by putting a piece of magnetized ferrite in the centre of a waveguide or microstrip junction. Figure 8.7 shows the arrangements of these two types. The junctions are symmetrical so, in the absence of the ferrite the input signal would excite equally the other two ports.

Consider first the waveguide junction shown in Fig. 8.7(a). An input

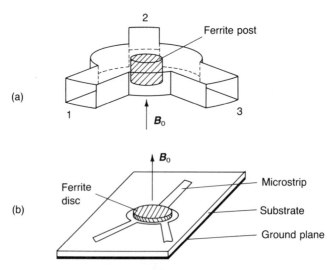

Fig. 8.7 Ferrite junction circulators: (a) waveguide, and (b) microstrip.

signal at port 1 can be thought of as coupling into the annular space around the ferrite rod. The symmetry of the system requires that a pair of waves of equal amplitude should be excited travelling round the circumference of the cavity in opposite directions. If the ferrite rod were replaced by a metal one then the phases of the signals excited at ports 2 and 3 would be the same. The presence of the ferrite alters the balance between the electrical lengths of the clockwise and counterclockwise paths. The dimensions of the cavity and of the ferrite and the strength of the magnetic field can be chosen so that the two waves are in antiphase at port 3. All the input power is then coupled out at port 2. Because of the symmetry of the device the same relationships hold for signals injected at ports 2 and 3 with cyclical permutation of the ports.

The microstrip junction circulator shown in Fig. 8.7(b) works in the same way as the waveguide one. An alternative view of the process is to consider the junction as a resonator. The size of the junction is chosen so that the TM_{011} resonance coincides with the centre frequency of operation. All possible orientations of this mode can be represented as sums of the two modes with null planes at right angles shown in Figs. 8.8(a) and (b). An input signal at port 1 only excites the first of these two modes. Now any

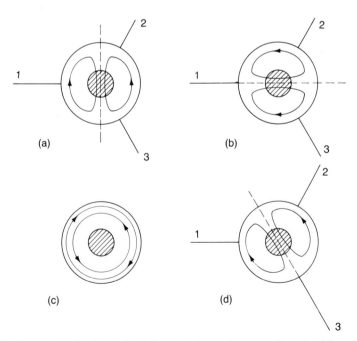

Fig. 8.8 Operation of a ferrite junction circulator: the normal modes (a) and (b) of the TM_{011} resonance of the junction can be thought of as the superposition of circulating waves (c). The propagation of these waves is affected differently by the ferrite for the two directions so that the null plane is rotated as shown in (d).

resonance can be thought of as the superposition of pairs of equal and opposite waves. In this case the waves are circumferential waves as shown in Fig. 8.8(c). The presence of the ferrite alters the propagation constants of these two waves so that their resonant frequencies differ from one another. If the operating frequency is chosen to lie between these two resonant frequencies the effect is that the two waves are excited with different phases. When they are superimposed the result is that the null plane is found to be rotated from its original position (Fig. 8.8(a)) somewhat as shown in Fig. 8.8(d). The dimensions of the ferrite and the strength of the field are chosen so that the null plane is aligned with port 3.

If one of the ports of a three-port junction circulator is terminated by a matched load the result is an isolator. Devices described as isolators are sometimes junction circulators with one port matched internally. High-power isolators, needed to protect the power amplifiers of transmitters from reflected signals, are made by terminating one of the ports of a circulator by a high-power matched load.

8.6 FARADAY ROTATION DEVICES

In Chapter 1 we saw that when a plane-polarized wave passes through a gyromagnetic material the plane of polarization is rotated. This is known as Faraday rotation. The sense of the rotation is independent of the direction of propagation of the signal as shown in Fig. 8.9. The axis of rotation coincides with the direction of the static magnetic field.

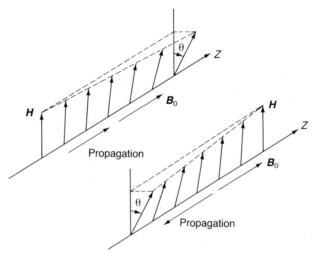

Fig. 8.9 Faraday rotation of the plane of polarization of a wave in a gyromagnetic material. The sense of rotation is the same for both directions of propagation.

Because Faraday rotation is a non-reciprocal phenomenon it can be used to make gyrators and isolators. Figure 8.10 shows the arrangement of a Faraday rotation gyrator. The dimensions of the device and the strength of the static field are chosen so that the plane of polarization of the wave is rotated through 90°. The figure shows how the plane of polarization is rotated for both forward and backward waves. It is evident that the difference between the phase shifts for the two waves is 180°. If it is inconvenient to have the input and output waveguides rotated with respect to each other a waveguide twist can be added to bring them back into line. Although the theory given in this book only covers the propagation of plane waves in infinite gyromagnetic madia it can be shown that Faraday rotation devices can work satisfactorily when the ferrite is in the form of a rod on the axis of the waveguide. A Faraday rotation gyrator can be used in place of a resonance gyrator to make a phase shift isolator.

An alternative type of isolator which makes use of Faraday rotation is shown in Fig. 8.11. The input and output wave guides are rotated by 45° with respect to one another and the central section introduces 45° of

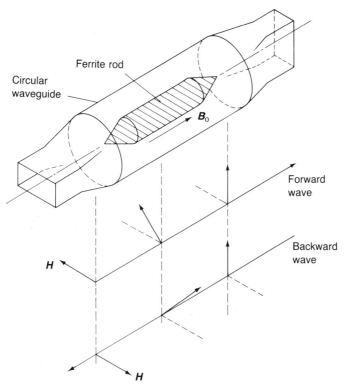

Fig. 8.10 (a) Arrangement of a Faraday rotation gyrator. The plane of polarization of a wave is rotated as it passes through the device is shown in (b).

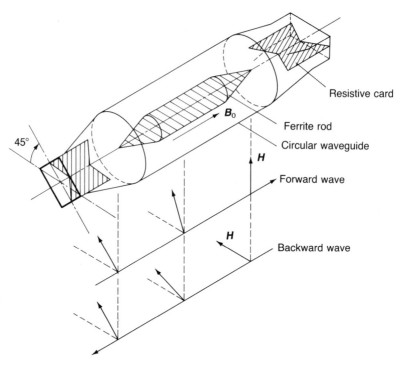

Fig. 8.11 Arrangement of a Faraday rotation isolator. Forward waves are absorbed by the resistive card at the output; backward waves travel through unimpeded.

Faraday rotation. The effects on forward and backward waves are shown in the diagram. The plane of polarization of the forward wave is rotated so that the magnetic field is parallel to the narrow walls of the guide at port 2. This mode is cut off and a vane of resistive card is arranged parallel to the electric field so that the power is absorbed and not reflected back through the device. A backward wave is rotated counterclockwise so that its magnetic field is parallel to the broad walls of the guide at port 1 and it therefore passes through the device with very little attenuation. The relative orientations of the guides can be restored by a 45° twist as before if necessary.

Faraday rotation devices are inherently narrow band because the rotation depends upon the electrical length of the ferrite. For this reason they are normally only used at millimetre wave frequencies where it is difficult to construct devices which rely on gyromagnetic resonance.

8.7 EDGE-MODE DEVICES

An interesting phenomenon occurs when a microstrip circuit is built on a ferrite substrate. It is found that the fields are concentrated under one edge

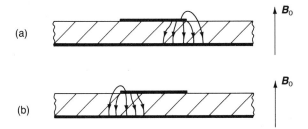

Fig. 8.12 Edge modes in a microstrip line on a ferrite substrate: (a) waves travelling into the paper, and (b) waves travelling out of the paper.

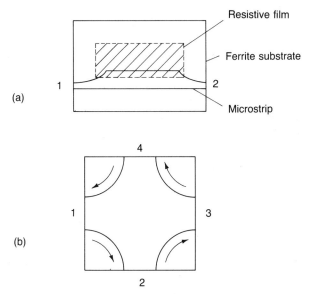

Fig. 8.13 Edge-mode devices: (a) edge-mode isolator, and (b) edge-mode circulator.

of the stripline depending upon the direction of propagation of the waves, as shown in Fig. 8.12. This effect can be used to make both isolators and circulators.

If the width of the microstrip is increased the edge modes tend to follow the edges. It is therefore possible to separate them. Figure 8.13(a) shows an edge-mode isolator. Waves travelling from port 1 to port 2 pass unattenuated whilst waves in the opposite direction pass into the region covered by the resistive film and are absorbed. This device can work over a very wide band (more than two octaves) because it does not depend upon resonance or phase shift for its operation. The insertion loss for backward waves can exceed 30 dB.

Figure 8.13(b) shows an edge-mode circulator. A signal injected at port

1 appears at port 2 and so on. Like the edge-mode isolator this device can work over a very broad band. There is also no reason why the number of ports should not be increased.

Edge mode devices suffer from the disadvantage that the loss is rather high in the forward direction compared with other kinds of device.

8.8 YIG FILTERS

In Chapter 7 we saw that resonators form the basis of microwave filters. The gyromagnetic resonance of ferrite materials can be used in the same way. Yttrium iron garnet (YIG) has a very narrow resonance so it is the usual choice for this purpose.

The basic element in a YIG filter is a sphere of the material which is coupled magnetically to two transmission lines as shown in Fig. 8.14. Because the two transmission lines are arranged at right angles to each other there is virtually no coupling between them except when the signal frequency is equal to the gyromagnetic resonant frequency. The resonance couples the two lines together because of the off-diagonal terms in the permeability tensor (8.18). These mean that a field in the x direction induces magnetisation of the sphere in the y direction as well as in the x direction.

The YIG element is made in the form of a sphere because it can be shown that a uniform external magnetic field produces a uniform magnetic field within it (Bleaney and Bleaney, 1976). That ensures that all the dipoles precess at the same frequency so that the resonance line is as sharp as possible. Bandwidths of around 20 MHz can be obtained over the range 500 MHz to 40 GHz with a centre frequency selected by the strength of the external magnetic field. This property of a YIG filter makes it especially useful for applications such as sweep oscillators, frequency counters and spectrum analysers because it can be tuned rapidly by electrical means.

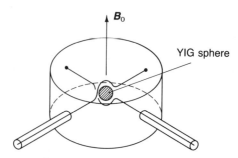

Fig. 8.14 Arrangement of a YIG filter.

8.9 CONCLUSION

In this chapter we have examined the microwave properties of ferrites and the ways in which they are exploited in a variety of devices.

Ferrites contain unpaired electrons whose magnetic dipole moments cause them to precess around the direction of an external magnetic field. The result of this precession is that the permeability of a magnetized ferrite is not a scalar as far as a.c. fields are concerned. There is coupling between the two field directions normal to the static magnetizing field which results in the Faraday rotation of plane-polarized waves. For circularly polarized waves the permeability is a scalar so these waves propagate through the ferrite in a stable manner. However it is found that the permeability is not the same for the two directions of circular polarization. In particular, for one of them, there is a resonance which produces strong losses.

The non-reciprocal properties of ferrites are put to use in isolators, circulators and filters. Isolators which allow a signal to pass unattenuated in only one direction make use of either the absorption peak or the non-reciprocal phase shift properties of a ferrite in a waveguide. Circulators in which a set of input and output ports are connected together in a cyclical manner can be made using non-reciprocal phase shifters. An alternative design which is particularly compact makes use of a ferrite loaded junction between waveguides or transmission lines.

Microstrip manufactured on a ferrite substrate shows the interesting property that the electric field of a wave is concentrated under one or the other edge of the stripline depending upon the direction of propagation. The effect is used to make a class of isolators and circulators employing edge modes.

Finally, the sharp resonance of yttrium iron garnet ferrite (YIG) is used to make electrically tunable filters with very narrow pass bands. These filters find application in a wide range of microwave instruments.

Further information on ferrites can be found in Baden-Fuller (1987).

EXERCISES

8.1 A ferrite material has a saturation magnetization of 0.36 T and a maximum relative permeability of 2000. Investigate the variation with frequency in the range 1 to 20 GHz of the real and imaginary parts of the permeability if the magnetizing field is 0.3 T. Assume that the damping constant $\alpha = 0.05$ and that $g = 2.0$.

8.2 Calculate the propagation and damping constants of the material in Question 8.1 over the same frequency range for positive and negative circularly polarized waves.

8.3 Investigate the solutions to equation (8.30) for frequencies in the range 8.0 to 12.0 GHz in WG16 waveguide.

<table>
<tr><td>9</td><td># Solid state microwave devices</td></tr>
</table>

9.1 INTRODUCTION

The devices considered so far in this book are passive linear devices. In order to make useful systems we also require active devices which can convert d.c. into r.f. energy (oscillators and amplifiers) and non-linear devices for mixing, detection and switching. There are two main groups of devices which can perform these functions: those which depend upon the movement of electrons in semiconductor materials (considered in this chapter) and those in which the electrons move in a vacuum or a gas discharge (considered in Chapter 10).

This book is about electromagnetic waves and their applications so we shall concentrate upon the terminal properties of the different kinds of semiconductor device and the way in which they are put to use. For information about the inner workings of the devices the reader must refer to other books. (e.g. Bar-Lev, 1979).

There are two main classes of semiconductor device: diodes (with two terminals) and transistors (with three). Devices designed to work at high frequencies differ in some respects from their low-frequency counterparts. High-frequency diodes and transistors are dicussed in Sections 9.3 and 9.4. In the rest of the chapter we shall see how their properties are employed to make a variety of useful subsystems.

9.2 SEMICONDUCTOR MATERIALS

The commonest semiconductor material is silicon. This element has four valence electrons and it forms crystals in which each atom has four nearest neighbours to which it is bound by covalent bonds. These bonds each involve two electrons: one from each atom.

In an isolated atom the electrons which surround the nucleus cannot have arbitrary energies. Instead there is a set of permitted energy states which they may occupy. According to the Pauli exclusion principle each state may only accommodate two electrons. When electrons move from one state to another they absorb or emit energy in the form of packets of electromagnetic radiation known as photons. The frequency of the photon emitted when an electron moves between two states with energies E_1 and E_2 is given by

$$f = (E_1 - E_2)/h, \qquad (9.1)$$

where h is Planck's constant which has the value 6.625×10^{-34} J s. The frequencies associated with transitions in isolated atoms normally fall in the optical region of the electromagnetic spectrum. They account for the characteristic colours of, for example, sodium and neon discharge lamps.

When atoms are assembled together to form a crystalline solid the individual energy levels are perturbed and form corresponding energy bands each containing $2N$ closely spaced energy states, where N is the number of atoms in the crystal. The energy differences between these levels are very small so that thermal energy is sufficient to produce transitions between them. The typical thermal energy available is given by

$$E = kT, \qquad (9.2)$$

where T is the absolute temperature and k is Boltzmann's constant which has the numerical value 1.380×10^{-23} J K^{-1}. Normally it is only the highest energy levels which play any part in determining the physical properties of the crystal.

In silicon there are two energy bands which are of interest. These are known as the conduction band and the valence band. They are shown diagrammatically in Fig. 9.1. At the absolute zero of temperature the valence band would be full of electrons and the conduction band empty. The two bands are separated by a forbidden region whose width is 1.11 electron volts. (One electron volt (eV) is the change in the energy of an electron when it moves through a potential difference of one volt). At room temperature a tiny proportion of the electrons in the valence band can aquire enough energy to excite them into states in the conduction

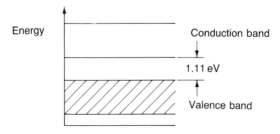

Fig. 9.1 Band diagram for silicon.

band. They are then free to change to other energy states near by, that is they are free to move through the crystal and form a conduction current. At the same time the states vacated at the top of the valence band allow some redistribution of the electrons there. This appears in experiments to be a motion of positively charged particles which are known as 'holes'. In pure silicon the result of thermal excitation is always to produce mobile charge carriers as electron–hole pairs. The thermal production of electron–hole pairs is balanced by recombination to produce a dynamic steady state. The number of charge carriers is only a tiny fraction of the total number of electrons in the valence band so the electrical conductivity of silicon is low. It is referred to as an intrinsic semiconducting material.

The properties of a semiconductor material can be altered by adding tiny quantities of other elements when the crystal is being grown or by subsequent diffusion or ion implantation. Usually these are chosen to have three or five valence electrons. Their presence produces additional holes or additional conduction electrons respectively so that there are no longer equal numbers of the two types of charge carrier. At room temperature virtually all the energy states associated with the impurity atoms are ionized. Material in which holes are the dominant charge carriers is known as p-type and that in which electrons dominate as n-type. The carrier densities and, hence, the electrical conductivity can be controlled by the impurity concentration. Material which has high conductivity is distinguished by the symbols p^+ and n^+.

Besides silicon there are other semiconductor materials. Germanium which was used to make the first bipolar transistors is still used for special purposes. More interesting at high frequencies are the intermetallic compounds, especially gallium arsenide (GaAs). Gallium has three valence electrons and arsenic has five so that together they can form crystals having the same structure as those of silicon. The intermetallic compounds add considerably to the range of semiconductor materials available. They all have different physical properties so it is possible to select the material which is best for a particular purpose.

The most important property as far as high-frequency devices are concerned is the mobility of the charge carriers. This is defined by the equation

$$\mu = v_d/E, \tag{9.3}$$

where v_d is the velocity with which the carriers drift through the material under the influence of an electric field E. The process occurring is that each electron is accelerated by the field only to have the energy gained turned into random thermal energy by a collision with an irregularity in the crystal structure. The process is repeated continually with the result that the whole assembly of charge carriers has a drift velocity superimposed upon a random thermal motion. The properties of some of the more common semiconductor materials are given in Table 9.1.

Table 9.1 Properties of semiconductor materials

	Si	Ge	GaAs	InSb
Energy gap (eV)	1.11	0.67	1.4	0.18
Gap type	Ind.	Ind.	Dir.	Dir.
Electron mobility $(cm^2\,V^{-1}\,s^{-1})$	1350	3900	8500	80 000
Hole mobility $(cm^2\,V^{-1}\,s^{-1})$	480	1900	450	150

Table 9.1 shows that GaAs has a much higher electron mobility than silicon. For this reason devices for high-speed and high-frequency operation are usually made from GaAs. Superficially InSb (indium antimonide) would be an even better choice. The reason why it and GaAs are not more generally used is that the technology needed to make devices is more difficult than that for silicon.

One other important property of semiconductor materials is shown in Table 9.1 as the gap type. This affects the way in which energy is released if an electron and a hole recombine. In direct band gap materials the energy is released as a photon whilst in indirect band gap materials it is released as thermal energy. That is why silicon and germanium cannot be used to make light-emitting diodes and semiconductor lasers.

9.3 DIODES

A semiconductor junction diode is formed by creating adjacent regions of p-type and n-type material as shown schematically in Fig. 9.2(a). These regions are characterized by having holes and electrons respectively as the dominant carrier types. It is found that there is a region at the junction where the carrier densities are much lower than in the bulk material on either side. This region is known as the depletion layer. If the diode is biased so that the p-type material is positive with respect to the n-type

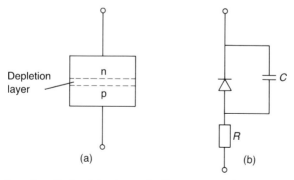

Fig. 9.2 p–n junction diode: (a) schematic diagram, and (b) equivalent circuit.

material majority charge carriers are injected across the junction and the diode conducts. If the opposite bias is applied the majority carrier cannot cross the junction and the diode carries only a very small current.

The $I-V$ characteristic of a semiconductor diode is given by

$$I = I_s \left[\exp \left(\frac{V}{\eta V_T} \right) - 1 \right], \tag{9.4}$$

where I_s is the reverse-bias saturation current, V_T (= kT/e) is the volt equivalent of temperature (25 mV at room temperature) and η is a constant whose value depends upon the material from which the diode is made.

The type of diode generally used at low frequencies is the silicon p–n junction shown in Fig. 9.2(a). This diode can be represented by the equivalent circuit of Fig. 9.2(b) in which the capacitance represents both the capacitance between the p and n regions and the storage of charge in the depletion region which separates them. The resistor represents the bulk resistance of the p and n regions. At high frequencies this kind of diode is unsatisfactory for two reasons. The first is the large junction capacitance which effectively short circuits the diode. The second is the slow speed of reaction caused by the low mobility of the charges and by storage of charge in the depletion layer. The capacitance of a reverse-biased junction varies with the bias voltage because the width of the depletion layer changes. This effect enables diodes to be used as voltage-variable capacitors. Diodes made for this purpose are known as varactor diodes, they are used to make electrically tunable circuits.

The problem of slow response can be overcome by using a metal–semiconductor junction in place of a p–n junction. Such a diode, known as a Schottky barrier diode, has a much faster switching time than a p–n junction, typically around 10 picoseconds. Figure 9.3 shows two examples of diodes of this type. In the point-contact diode (Fig. 9.3(a)) a metal whisker makes contact with the surface of a piece of semiconductor material. The usual combinations are p-type silicon with tungsten and n-type germanium with titanium. It important that the contact area between the two materials is kept as small as possible to minimize the junction capacitance. The other contact is made via metallization of the semiconductor. It is important that this is an 'ohmic' (non-rectifying) contact. The point-contact diode traces its ancestry to the 'cat's whisker' detectors used in early radio sets. It is now obsolescent because of its mechanical and electrical fragility.

Developments in semiconductor materials and technology have led to the form of Schottky barrier diode shown in Fig. 9.3(b). A metal contact makes contact with a semiconductor layer through a hole etched in an insulating oxide layer. The semiconductor layer is backed by a heavily doped, high-conductivity, layer of the same type of semiconductor and

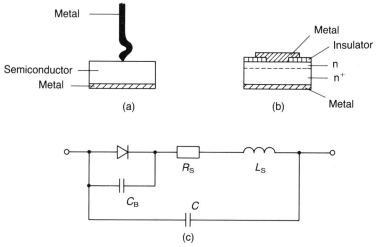

Fig. 9.3 Schottky barrier diodes: (a) point-contact diode, (b) modern form of Schottky barrier diode, and (c) equivalent circuit.

this, in turn, makes an ohmic contact with a metal base. The diameter of the contact between the metal and the semiconductor can be as small as $2\,\mu m$. The metal is commonly nickel, titanium or gold. The semiconductor is p-type or n-type silicon or n-type gallium arsenide. The latter has higher carrier mobilities than silicon and is therefore used to make diodes for millimetre wave frequencies (above 30 GHz).

The diodes shown in Figs. 9.3(a) and (b) can be represented by the equivalent circuit of Fig. 9.3(c) in which C_B is the capacitance of the junction, R_S the series resistance and L_S and C are the inductance and capacitance associated with the packaging of the diode. The I–V characteristic is given by the diode equation (9.4). For a silicon junction diode η is around 1.3 whilst for a silicon Schottky barrier diode it is typically 1.03 giving a faster turn on as the forward voltage rises. V in (9.4) is the voltage across the junction which is less than the terminal voltage by the voltage drop in R_S. At low currents the difference is negligible. The I–V characteristic curve (Fig. 9.4) has the same features as those of other semiconductor diodes with low forward resistance, high reverse resistance and breakdown at some reverse voltage.

The tunnel diode, first described by Esaki in 1958, is a p–n junction diode in which the two regions are very heavily doped and the depletion layer between them is very thin. Classically this layer is a barrier to the movement of charge across the junction but quantum theory shows that an appreciable number of electrons can 'tunnel' through it. Under conditions of zero bias the maximum electron energies on the two sides of the junction are equal and no net current flows. As forward bias is applied it is found

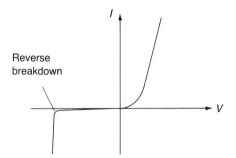

Fig. 9.4 Typical characteristic curve for a semiconductor diode.

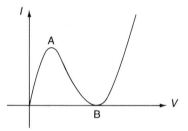

Fig. 9.5 Characteristic curve for a tunnel diode.

that the current increases to a maximum and then falls off again. For still higher forward bias the ordinary injection current of a p–n junction starts to flow and the diode conducts. Figure 9.5 shows the resulting $I-V$ characteristic. The importance of this curve is that it has a region A–B in which the dynamic resistance is negative. This negative resistance can be used to generate oscillations as we shall see in Section 9.7.

The reverse breakdown condition of a semiconductor diode shown in Fig. 9.4 can be caused by tunnelling but is more commonly the result of avalanche breakdown. Ordinarily the charge carriers moving through a semiconductor material are accelerated by the applied electric field only to have their velocity randomized after a short while by collisions with lattice defects. The result is that a net drift velocity is superimposed upon the random thermal velocities. However, if the electric field is strong enough, the charge carriers may gain sufficient energy to cause ionization when they make collisions. The result is the creation of an additional pair of charge carriers (electron and hole) which can in their turn make ionizing collisions. This phenomenon is like an uncontrolled chain reaction producing a rapid increase in the number of charge carriers and in the current flowing through the device. Avalanche effects are used to produce negative resistance in IMPATT and TRAPATT diodes.

The IMPATT (IMPact Avalanche Transit Time) diode was proposed by

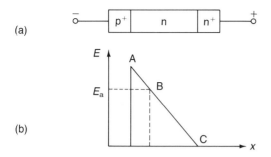

Fig. 9.6 IMPATT diode: (a) schematic diagram, and (b) electric field distribution within the device.

Read in 1958 and first demonstrated experimentally in 1965. A typical IMPATT idode has the structure shown in Fig. 9.6(a). The layers are, from left to right, a heavily doped p region, a long lightly doped n region and a heavily doped n region. When this diode is strongly reverse biased the region A–C is depleted of free charges and contains a strong electric field since most of the external potential difference appears across it. The resulting field distribution is shown in Fig. 9.6(b). If the external field is strong enough an avalanche builds up in A–B and the electrons produced move to B and drift across B–C. Now avalanching depends critically upon the applied field so, if a d.c. bias is applied to a level just below that at which an avalanche will develop then the superposition of a small a.c. signal causes avalanching for part of each a.c. cycle. Pulses of current are then injected into B–C. It is a property of semiconductor materials that, at high field strengths, the drift velocity tends to a constant value. The time taken for the current pulses to drift across B–C is then independent of the external field. If the length B–C is chosen correctly the current pulses emerge at the terminals of the device in antiphase with the applied a.c. signal. The device is then exhibiting a negative dynamic resistance and it may be used as the basis of an oscillator. The TRAPATT (TRApped Plasma Avalanche Triggered Transit) diode which has a similar structure to the IMPATT diode also shows negative dynamic resistance. Details of its mode of operation are given by Liao (1980).

The Gunn diode or transferred electron device depends for its operation upon the special properties of gallium arsenide. This material has two energy bands which can contain conduction electrons. One of these bands is at a lower energy level than the other and has the greater value of electron mobility. If a sample of GaAs has a voltage across it some of the electrons gain energy and are transferred from the lower energy band to the higher one. They then have lower mobilities and contribute less to the current flow through the material. This state is unstable and space-charge instability domains form which travel through the material producing an

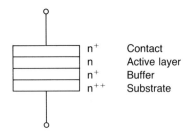

Fig. 9.7 Schematic diagram of a Gunn diode.

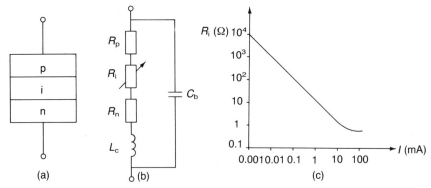

Fig. 9.8 PIN diode: (a) schematic diagram, (b) equivalent circuit, and (c) variation of R_i with current.

a.c. current which is superimposed upon the d.c. bias current. This effect was discovered experimentally by Gunn in 1963. It had been predicted by Ridley, Watkins and Hilsum a few years earlier but, at that time, no GaAs of sufficient quality was available for its experimental demonstration. Gunn diodes do not depend upon p–n junctions for their operation. A typical device structure is shown in Fig. 9.7.

The final example of a diode with microwave applications is the PIN diode shown in Fig. 9.8(a). In this device a layer of intrinsic semiconductor material is sandwiched between p-type and n-type layers. When the device is reverse biased it behaves as a capacitor and no current flows. When it is forward biased charge carriers are injected into the intrinsic region at both junctions so that the region conducts. It is found that the resistance of the intrinsic region can be varied from close to zero to almost infinity by varying the bias current. The PIN diode can therefore be used as a variable attenuator or a switch by connecting it across a transmission line or waveguide. The equivalent circuit of a forward-biased PIN diode is shown in Fig. 9.8(b). Here R_p, R_i and R_n are the resistances of the layers of the diode, L_c is the inductance of the conductors and C_b is the parasitic capacitance of the encapsualtion. The variation of R_i with bias current is shown in Fig. 9.8(c).

9.4 TRANSISTORS

Bipolar and field-effect transistors are familiar circuit elements at low frequencies. High-frequency versions of both these devices exist but at frequencies above 4 GHz the field-effect transistor is much better than its bipolar rival.

Figure 9.9(a) shows the arrangement of a typical n–p–n silicon planar bipolar transistor. At high frequencies it can be represented by the small-signal hybrid-pi equivalent circuit shown in Fig. 9.9(b). The derivation of this model and the physical significance of the components are discussed by Bar-Lev (1979). It is shown in that book that the common-emitter short-circuit current gain obeys the equation

$$\frac{\beta}{\beta_0} = \frac{1}{1 + j(\omega/\omega_\beta)},\tag{9.5}$$

where β_0 is the low-frequency value of β and the corner frequency is given by

$$\omega_\beta = \frac{1}{\beta_0(C_e + C_c + C_{\text{diff}})r_e}.\tag{9.6}$$

The bandwidth of the transistor can be defined by the frequency at which $|\beta| = 1$. If this frequency is large compared with ω_β then it is given with sufficient accuracy by

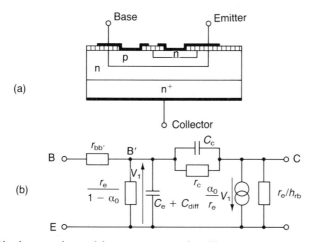

Fig. 9.9 Bipolar transistor: (a) arrangement of a silicon planar n–p–n transistor, and (b) the hybrid pi equivalent circuit.

$$\frac{1}{\omega_T} = \frac{1}{\beta_0 \omega_\beta} = r_e(C_e + C_c + C_{diff}).$$ (9.7)

It can also be shown that the power gain of the transistor working into a matched load is

$$G = \frac{\omega_T}{4\omega^2 r_{bb'} C_c}.$$ (9.8)

To obtain the best high-frequency performance we must therefore minimize $r_{bb'}$ and C_c and maximize ω_T.

The base spreading resistance $r_{bb'}$ is the resistance of the thin base layer between the base terminal and the active part of the transistor. C_c represents the capacitance of the reverse-biased collector–base junction. Equation (9.7) does not include all the factors affecting the bandwidth of a transistor at high frequencies. Other factors include the transit times across the collector junction and the base and the effect of the series resistance of the collector on the charging time of C_c.

Thus, in order to make a high-frequency transistor, a number of steps can be taken.

1. The geometry of the transistor is made as small as possible to minimize the junction capacitances.
2. The thickness of the emitter is made as small as possible and the emitter–base junction is operated at the highest possible forward bias to minimize r_e.
3. The base is made as thin as possible and the doping graded to provide an internal electric field to minimize the base transit time.
4. The transistor is made in the interdigital form shown in Fig. 9.10 to minimize the base spreading resistance $r_{bb'}$.

The other thing which can be done to improve the high-frequency performance of a bipolar transistor is to used a material other than silicon. Gallium arsenide (GaAs) is one possibility because the electron mobility is higher than in silicon. It also has the advantage that it can work at higher

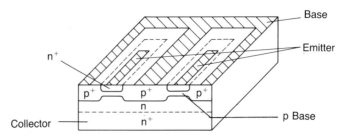

Fig. 9.10 Interdigital structure used for high-frequency transistors.

temperatures so allowing more power to be handled within the small sizes which must be used at high frequencies. Recent research has concentrated on the development of heterojunction transistors in which the junctions are between different semiconductor materials (e.g. gallium aluminium arsenide (GaAlAs) and GaAs). Bipolar transistors with cut-off frequencies of 40 GHz have been made in this way.

Gallium arsenide is used to make microwave field-effect transistors. These are akin to the low-frequency junction field-effect transistor (JFET) but employ a metal semiconductor Schottky junction in place of a p–n junction. This device which is known as a MESFET (MEtal Semiconductor FET) has the arrangement shown in Fig. 9.11. The active layer of high-conductivity n-type GaAs less than a micrometre in thickness is produced by ion implantation or epitaxial growth upon a semi-insulating substrate. A high-resistivity buffer layer separates the two. The source and drain contacts are ohmic (non-rectifying) whilst that of the gate forms a Schottky diode. The Schottky junction produces a region in the active layer which is depleted of conduction electrons and whose thickness is controlled by the reverse bias applied to the junction. Thus the thickness of the conducting channel, and therefore its resistance, is controlled by the source–gate voltage. The separation of the source and the drain is typically a few micrometres whilst the lengths of the electrodes are perhaps a hundred micrometres.

A simple equivalent circuit of the MESFET can be constructed by considering Fig. 9.11 with the result shown in Fig. 9.12(a). In this circuit R_I takes account of the potential difference between the source and the active region of the channel under the gate. R_{DS} is the channel resistance and the transconductance g_m represents the small-signal effect of the gate-channel voltage on the current flowing between source and drain. This equivalent circuit is not capable of representing the transistor correctly at high frequencies. For that purpose a much more complicated circuit is used which includes the parasitic capacitances and inductances associated with the packaging of the device with the result shown in Fig. 9.12(b). Circuits like this one are used to represent MESFETs in computer-aided-design programs for microwave circuits.

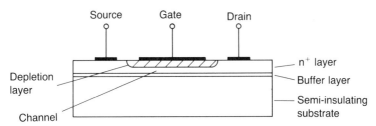

Fig. 9.11 Arrangement of a metal semiconductor field-effect transistor (MESFET).

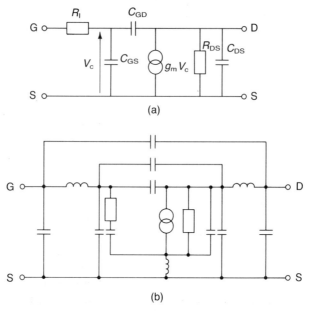

Fig. 9.12 Equivalent circuits for MESFETs: (a) basic equivalent circuit for the device, and (b) equivalent circuit which includes the parasitic effects associated with packaging.

9.5 DETECTORS AND MIXERS

The non-linear properties of diodes find application in detectors and mixers. The diode equation (9.4) can be expanded as a power series to give

$$I = I_s(\alpha V + \tfrac{1}{2}\alpha^2 V^2 + \cdots), \tag{9.9}$$

where $\alpha = 1/\eta V_T$. For small signals the higher-order terms can be neglected. If the voltage applied to the diode varies with time as $V_0 \cos \omega t$ the current is

$$I = I_s[\alpha V_0 \cos \omega t + \tfrac{1}{2}\alpha^2 V_0^2(\tfrac{1}{2} + \tfrac{1}{2}\cos 2\omega t)] \tag{9.10}$$

if terms beyond the second are neglected. The equivalent circuit of the diode (Fig. 9.3(c)) has shunt capacitance and series inductance and therefore acts as a low-pass filter. External components can be added if necessary to adjust the cut-off frequency so that it lies below the signal frequency ω. The current flowing in the external circuit is then just the d.c. component of (9.10), namely

$$I = \tfrac{1}{4}I_s\alpha V_0^2. \tag{9.11}$$

There is thus a square-law relationship between the applied voltage and the current flow in the diode, and measuring instruments such as VSWR

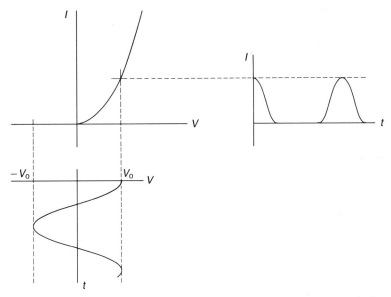

Fig. 9.13 Use of a diode as a detector showing how current flows only during the positive half cycles of the voltage waveform.

meters are calibrated on this basis. The rectifying operation of a diode can be represented graphically as shown in Fig. 9.13. The graph below the diode characteristic curve shows the variation of the signal voltage with time. When V is positive the diode conducts whereas in the negative half cycle virtually no current flows. The resulting current waveform is shown on the right of the diagram. The time average current flow is evidently non-zero.

In microwave measuring systems the signal is commonly square-wave modulated in order to improve the signal-to-noise ratio of the system. Provided that the modulating frequency is below the cut-off of the low-pass filter in the diode circuit the output current waveform is a square wave at the modulating frequency.

Detector diodes are often mounted in coaxial form as shown in Fig. 9.14(a). They can also be mounted directly in a waveguide as shown in Fig. 9.14(b) and this is a common configuration in detectors for radar receivers. For laboratory instrumentation it is more usual to use a coaxial detector mount attached to a waveguide–coaxial line transformer.

If a diode is subjected simultaneously to two signals at different frequencies the result is the production of signals at other frequencies because of the non-linearity of the diode. Suppose that the input signal is

$$V = V_1 \cos \omega_1 t + V_2 \cos \omega_2 t \qquad (9.12)$$

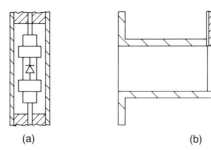

Fig. 9.14 Microwave diode mounts: (a) coaxial line, and (b) waveguide.

then, substituting into (9.9) we obtain

$$I = I_s[\alpha V_1 \cos \omega_1 t + \alpha V_2 \cos \omega_2 t + \tfrac{1}{2}\alpha^2 V_1^2 \cos^2 \omega_1 t$$
$$+ \tfrac{1}{2}\alpha^2 V_2^2 \cos^2 \omega_2 t + \alpha^2 V_1 V_2 \cos \omega_1 t \cos \omega_2 t]$$
$$= I_s[\alpha V_1 \cos \omega_1 t + \alpha V_2 \cos \omega_2 t + \tfrac{1}{4}\alpha^2 (V_1^2 + V_2^2)$$
$$+ \tfrac{1}{4}\alpha^2 V_1^2 \cos 2\omega_1 t + \tfrac{1}{4}\alpha^2 V_2^2 \cos 2\omega_2 t$$
$$+ \tfrac{1}{2}\alpha^2 V_1 V_2 \cos (\omega_1 - \omega_2)t + \tfrac{1}{2}\alpha^2 V_1 V_2 \cos (\omega_1 + \omega_2)t] \qquad (9.13)$$

which contains signals at frequencies ω_1, ω_2, $2\omega_1$, $2\omega_2$, $(\omega_1 - \omega_2)$ and $(\omega_1 + \omega_2)$. If the higher-order terms are taken into account as well the resulting frequencies are

$$\omega = (m\omega_1 + n\omega_2), \qquad (9.14)$$

where $m, n = \pm 0, 1, 2$, etc. These frequencies are known as intermodulation products. In an amplifier they are undesirable because they produce distortion of the amplified signal but they can also be used for frequency changing and modulation. The simplest form of mixer circuit is the single-ended mixer shown in Fig. 9.15(a). The signals to be mixed are fed into the main arm and the side arm of a directional coupler and the output applied to a diode. A filter is then used to select whichever of the intermodulation products is required. Typical applications include conversion of a received r.f. signal to an intermediate frequency (i.f.) for further amplification and generation of a frequency-modulated carrier.

An improved form of mixer is the balanced mixer shown in Fig. 9.15(b). This is commonly used for the conversion of r.f. to i.f. In that case ω_1 is the frequency of the incoming r.f. signal and ω_2 that of the local oscillator. The local oscillator will have some noise output at the intermediate frequency and this limits the sensitivity of the receiver. If a hybrid junction such as a magic tee is used as the 3 dB coupler then the contributions of the local oscillator noise to the outputs of the two diodes are in antiphase so that they cancel each other out.

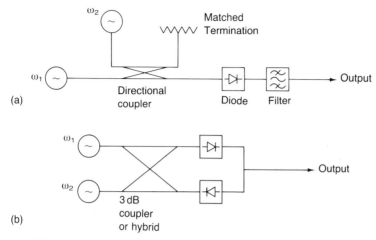

(a)

(b)

Fig. 9.15 Mixer circuits: (a) single-ended mixer, and (b) balanced mixer.

The circuits shown in Fig. 9.15 can be realized in microstrip, waveguide and other types of transmission line such as co-planar waveguide. For further information on mixers and their applications consult texts on telecommunications (Dunlop and Smith, 1984).

It is, of course, important that a diode used as a mixer or detector should be matched as far as possible to the transmission line in which it is mounted. The equivalent circuit of an encapsulated diode shown in Fig. 9.3(c) shows that a diode has both series and parallel resonances. At low frequencies the diode impedance is the forward resistance R_s. This is generally rather small. At the parallel resonant frequency the impedance is real and quite close to R_s. The series resonance on the other hand produces a high impedance which is generally closer to that of the transmission line.

9.6 SWITCHES

The PIN diode described in Section 9.3 can be used as a swtich. In its simplest form this device is a single diode which is arranged either in parallel or in series in a transmission line as shown in Fig. 9.16(a) and (b). In microstrip the series connection is easiest to realise whilst shunt connection must be used in waveguide. Because of the parasitic inductance and capacitance associated with the encapsulation of a diode the best results are obtained with unencapsulated diodes connected directly into a microstrip circuit. To obtain greater isolation it is possible to use several diodes. Figure 9.16(c) shows how three diodes could be arranged in waveguide so that their 'on' admittances add to give greater isolation. A switch of this kind can be used to provide square-wave modulation at high frequencies.

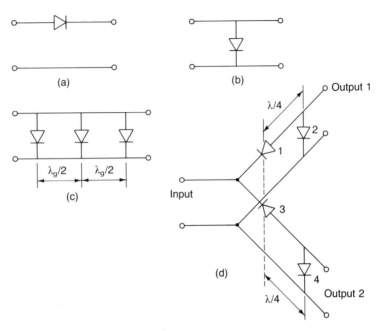

Fig. 9.16 PIN diode switching circuits: (a) series connection, (b) shunt connection, (c) use of several diodes to obtain greater isolation, and (d) arrangement of diodes to switch the input signal to either of two output ports.

Two or more switches can be combined to enable an input signal to be switched to one of two, or more, output lines. Figure 9.16(d) shows such an arrangement. The use of series and parallel diodes in pairs increases the isolation produced when a switch is open. When diodes 1 and 4 are 'on' and 2 and 3 are 'off' the output is directed to output port 1. Reversing the situation switches the output to port 2. The method of making the d.c. connections to the diodes is not shown. It requires a combination of capacitors and chokes to block d.c. and a.c. paths respectively.

Switches like that shown in Fig. 9.16(d) can be used to make switchable phase shifters. One method is to use pairs of switches to select alternative transmission lines with different lengths as shown in Fig. 9.17. By cascading

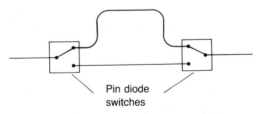

Fig. 9.17 Switchable phase shifter using PIN diode switches.

several switchable phase shifters it is possible to make an electrically variable phase shifter whose phase can be varied in small steps. Phase shifters
of this kind find application in circuits for feeding phased array antennas.

9.7 OSCILLATORS

An oscillator is a subsystem which converts d.c. energy into r.f. energy at a
specified frequency. The frequency is commonly selected by a parallel
resonant circuit as shown in Fig. 9.18. In order to make an oscillator some
kind of active device must be connected in parallel with the resonator to
supply the r.f. output power. At all frequencies except those close to the
resonant frequency the resonator has a high admittance and the output is
zero. At resonance the resistance R represents the losses in the resonator.
If the resonator were excited by a voltage pulse the oscillations would die
away after a few cycles because of these losses. Even if the resonator were
lossless so that the oscillations continued undamped it would not be possible
to extract any power from it without causing damping. It follows that the
condition for continuous oscillations to occur with useful output power is
for the impedance presented at the terminals A–B to be a negative resistance at the desired frequency. There are a number of ways in which this
negative resistance can be achieved.

In Section 9.3 we saw that tunnel diodes and IMPATT diodes exhibit
negative resistance characteristics. Tunnel diodes are very fragile and are
not in common use. IMPATT diodes, on the other hand, are valuable
because they are able to produce powers of a few watts of continuous-
wave (c.w.) power and several tens of watts of pulsed power at microwave
frequencies. Gunn diodes can operate in a number of modes but the one
normally used is the accumulation layer mode. In this mode the transit
time of charge carriers through the device is important and it can be thought
of as having negative dynamic resistance.

The performance of an oscillator of this kind depends critically upon the
Q factor of the resonator. A low Q factor gives poor frequency stability
and a higher noise figure. Oscillators commonly make use of waveguide,
coaxial line or microstrip resonators. The unloaded Q factor of a metal
waveguide resonator is typically 500 to 2000. Figure 9.19 shows the arrangement of a waveguide oscillator. The diode is placed approximately half a

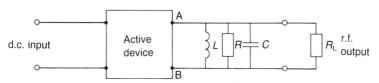

Fig. 9.18 General arrangement of an oscillator.

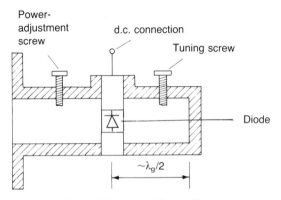

Fig. 9.19 Arrangement of a typical waveguide oscillator.

guide wavelength from a short circuit where the shunt impedance of the resonator is low and matched to that of the diode.

If a coaxial-line resonator is used the configuration is as shown in Fig. 9.20(a). At the resonant frequency the line is close to half a wavelength long, the difference being accounted for by the loading effect of the diode and its packaging. The unloaded Q factor is quite low (20 to 50) so the oscillator has poor stability and noise figure. This kind of arrangement can also include a varactor diode to allow the frequency to be changed electrically (Fig. 9.20(b)). Its main practical use is for testing diodes because of the ease with which matching can be achieved using coaxial line techniques.

Microstrip circuits tend to have high losses and therefore low Q factors. They are, however, simple, compact and cheap to manufacture. The frequency stability can be improved by using a separate high-Q resonator such as a YIG or dielectric resonator. Tuning can be achieved with YIG or with a varactor diode. YIG tuning provides bandwidths in excess of an octave and good temperature stability at the price of a relatively slow rate at which the frequency can be swept (the 'slew rate'). Varactor-tuned oscillators have narrower bandwidths and poorer temperature stability but higher slew rates. Table 9.2 shows a comparison between these two methods of tuning.

Table 9.2 Characteristics of voltage-tunable microwave oscillators

	YIG	Varactor
Bandwidth	Wide	Narrow
Q factor	High	Low
Linearity	$<\pm1\%$	$<\pm12\%$
Slew rate	$<1\,\mathrm{MHz\,s^{-1}}$	1 to $10\,\mathrm{GHz\,s^{-1}}$
Temperature frequency sensitivity	$<100\,\mathrm{p.p.m.\,K^{-1}}$	$<300\,\mathrm{p.p.m.\,K^{-1}}$

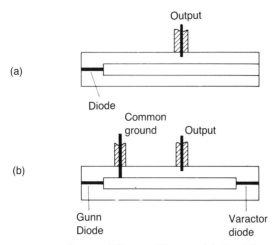

Fig. 9.20 Arrangements of coaxial-line oscillators: (a) fixed-frequency, and (b) varactor-diode tuned oscillators.

An alternative approach to oscillator design employs a three terminal active device, normally a MESFET. At frequencies up to 5 GHz good power output can be obtained using silicon bipolar junction transistors. At higher frequencies gallium arsenide FETs are used. Transistor oscillators tend to be noisier than those which use diodes but they have the advantage that they can be tuned over multioctave bands. This is useful in test equipment applications. The tuning ranges of diode oscillators are limited to around an octave by the transit time of electrons through the device.

9.8 AMPLIFIERS

The oscillators considered in the previous section are sources of microwave power whose frequency is controlled by the resonant circuit incorporated in them. An amplifier, in contrast, produces an output whose power is taken from the d.c. power supply but whose amplitude and frequency are controlled by the input signal.

Transistor amplifiers at microwave frequencies have configurations very like those familiar from low-frequency practice as shown in Fig. 9.21. There are, however, some differences.

1. The parasitic reactances of the devices are much more important than at low frequencies.
2. Networks must be designed to match the input and output to the external transmission lines.

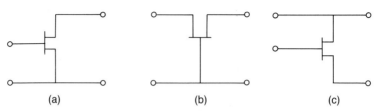

Fig. 9.21 FET amplifier configurations: (a) common source, (b) common gate, and (c) common drain.

3. Chokes and d.c. blocking capacitors must be used to separate the bias circuits from the r.f. circuits.

These factors make the design of microwave amplifiers much more difficult than the design of low-frequency amplifiers (Pengelly, 1986). Computer-aided design methods are normally employed for this purpose and the circuits realized using microstrip technology.

Microwave transistor amplifiers can readily be made with bandwidths of an octave. For a single stage the gain is low (a few decibels) and the power-added efficiency around 20 to 30%. As at low frequencies, it is possible to get increased bandwidth by using feedback to trade gain for bandwidth. Amplifiers have been made in this way with bandwidths of 3.5 octaves for use in electronic counter-measures (ECM) systems (Section 12.7). A second way of getting very wide bandwidth is to use a distributed amplifier circuit. Figure 9.22 shows the arrangement of such an amplifier with the bias connection omitted. Essentially it consists of a pair of transmission lines coupled together at intervals by transistors. The parasitic capacitances of the gates and drains modify the characteristics of the transmission lines but do not otherwise limit the performance of the devices. Multi-octave bandwidths can be achieved using this technique.

The power output available from a single transistor goes down as the frequency goes up. This is because transistors for high-frequency operation are smaller and so less able to dissipate heat. Powers of several watts at 10 GHz can be achieved in single devices. To obtain higher powers the outputs from several transistors can be combined using power dividers or

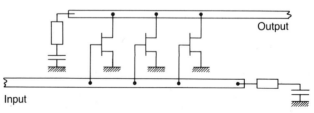

Fig. 9.22 Arrangement of a distributed amplifier.

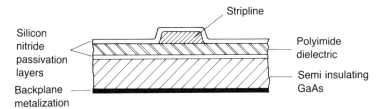

Fig. 9.23 Arrangement of stripline used in GaAs monolithic microwave integrated circuits (MMICs).

3 dB hybrid junctions operated in reverse. The insertion losses in these devices reduce the overall efficiency of the amplifier. A typical conversion efficiency for a microwave amplifier at 8 GHz is 15%.

9.9 MONOLITHIC MICROWAVE INTEGRATED CIRCUITS

Since 1960 the electronics industry has been dramatically altered by the development of silicon integrated circuits. The development of monolithic integrated circuits for use at microwave frequencies was much more difficult and circuits of this kind have only been available commercially since 1986. Monolithic microwave integrated circuits (MMICs) are constructed on gallium arsenide substrates using a miniaturized form of microstrip circuit. Figure 9.23 shows the construction of a stripline for a GaAs MMIC, the strip thickness is typically a few micrometres and its width a few tens of micrometres. The transistors may have gate lengths down to a few tenths of a micrometre and gate widths of a few hundred micrometres. One interesting effect of this miniaturization of circuits is that the dimensions are small compared with the free-space wavelength at microwave frequencies. It is therefore sometimes possible to use lumped-element circuit designs in MMIC technology.

A wide range of circuits is now available including amplifiers, oscillators, phase shifters, switches, mixers, and even complete receivers. Power outputs of up to a watt have been achieved at frequencies up to 18 GHz. It is to be expected that MMIC technology will have an effect on microwave engineering comparable with that of silicon technology at lower frequencies.

9.10 LEDS AND LASER DIODES

The devices discussed so far in this chapter have been for use at microwave frequencies. Semiconductor devices can also be used as sources at optical frequencies.

We have already noted that in direct band gap materials such as GaAs a photon is emitted when an electron and a hole recombine. When a p–n junction diode is forward biased majority carriers are injected across the junction into a region where they swell the population of minority carriers. The majority and minority population densities are then out of equilibrium with each other and recombination takes place. Light-emitting diodes (LEDs) which work on this principle are in everyday use as indicators and as elements in alpha-numeric displays.

For optical fibre communication systems it is desirable to have sources which are more intense than LEDs and which emit coherent radiation. Such a device is the laser. Here we shall discuss the laser (light amplification by stimulated emission of radiation) diode as an illustration of the principles upon which all lasers work.

When light interacts with matter one or more of three processes occur.

1. Absorption: a photon is absorbed and an electron excited to a higher energy state.
2. Spontaneous emission: an electron moves to a lower energy state and a photon is emitted.
3. Stimulated emission: a photon whose energy matches that of a transition stimulates an electron to undergo that transition emitting another photon of the same frequency and phase as the first.

Normally, higher energy states are less densely populated than lower energy states and a photon is much more likely to be absorbed than to cause stimulated emission. The population distribution is determined by considerations of thermal equilibrium. It also depends upon the average time an electron will spend in a higher state before spontaneous emission occurs. In order for stimulated emission of radiation to become the dominant process two conditions must be satisfied:

1. there must be a high-energy state with a relatively long lifetime to which electrons can be excited, and
2. some way must be found of exciting enough electrons into this state to make stimulated emission more probable than absorption.

These conditions are satisfied for all lasers. They differ only in the mechanisms by which the second condition (population inversion) is brought about.

The final thing needed to make a laser work is some form of optical resonator to ensure a high density of photons at the frequency required for laser action within the device.

Figure 9.24 shows the arrangement of a semiconductor injection laser. The junction is arranged to be in a certain orientation with respect to the crystal structure of the semiconductor. This makes it possible for two opposing faces A–A to be cleaved accurately flat and parallel forming a Fabry–Pérot resonator. The p and n regions of the diode are heavily doped so that when a large forward current is passed through the diode large numbers of electrons are injected into the p region and large numbers

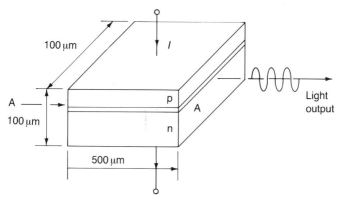

Fig. 9.24 Arrangement of a semiconductor injection laser.

of holes are injected into the n region. The diode is made from a direct band gap material so that the energy of recombination appears as photons. These travel through the diode stimulating more emission at the same frequency and phase. The light is partially reflected at the faces A–A ensuring a high intensity of electromagnetic radiation within the diode. The remainder of the light is emitted as an intense coherent beam which may be used to carry a signal along an optical fibre. Laser diodes are made in a variety of intermetallic compounds. The output power is limited by the maximum current which can be passed through the diode without destroying it. Continuous output powers of up to 70 W have been obtained with driving currents up to 250 A.

9.11 CONCLUSION

In this chapter we have considered the types of solid state device which are used to generate or amplify signals at microwave and optical frequencies. The bipolar transistor which is the commonest device at low frequencies is only usable up to about 4 GHz. Above that frequency gallium arsenide MESFETs are used in amplifiers and oscillators. Several types of diode exist having negative resistance characteristics which can be used to sustain the oscillations in a resonant circuit. It is easiest to make high Q circuits using hollow metal waveguides and that technology is commonly used for oscillators when high stability is required. Microstrip circuits have much lower Q factors but good stability can be achieved by using them in conjunction with high Q YIG or dielectric resonators. Electronically tuned oscillators can be made using YIG or varactor diodes.

Infrared and optical signals can be generated using light-emitting diodes and lasers. The semiconductor injection laser was used to explain the principles of laser action.

Vacuum devices

10.1 INTRODUCTION

In this chapter we consider the interactions which occur between electro-magnetic waves and electrons moving in a vacuum. These phenomena have important applications in the generation of high-power radiation at micro-wave frequencies and above. There is not enough space in this book to discuss the technical details of these devices. The emphasis is on the funda-mental processes involved. For greater detail reference should be made to the books and papers listed in the bibliography.

10.2 SPACE-CHARGE WAVES

Many high-power microwave amplifiers depend for their operation upon linear electron beams which are constrained to pass along their axes as shown in Fig. 10.1. The amplification of the devices depends upon the exchange of energy between the electron beam and the electric fields of microwave circuits surrounding it. The equations which govern the motion of the electrons are, first, the continuity equation (1.15)

$$\nabla \cdot J = -\frac{\partial \varrho}{\partial t},\tag{10.1}$$

where J is the current density and ϱ the charge density; and second, Newton's second law of motion

$$\frac{\mathrm{d}v}{\mathrm{d}t} = \eta(E_\mathrm{c} + E_\mathrm{sc}),\tag{10.2}$$

where η is the charge-to-mass ratio of the electron and E_c and E_sc are the electric field applied to the beam and the field arising from the space charge of the electrons. Next we note that the current is related to the charge density and the velocity by

$$J = \varrho v\tag{10.3}$$

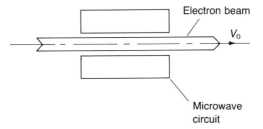

Fig. 10.1 General arrangement of a microwave linear beam tube.

and, finally, the relationship between the charge density and the space-charge field are given by Poisson's equation

$$\nabla \cdot \boldsymbol{E}_{\mathrm{sc}} = \varrho/\varepsilon_0. \tag{10.4}$$

Note that these four equations involve four dependent variables so, in principle, they can be solved. As a first stage of simplification let us assume that the electrons are moving parallel to the z axis and that the beam extends to infinity in the transverse direction. We will also assume that all quantities vary only with z and with time. The four equations then become

Continuity
$$\frac{\partial J}{\partial z} = -\frac{\partial \varrho}{\partial t}, \tag{10.5}$$

Motion
$$\frac{\mathrm{d}v}{\mathrm{d}t} = \frac{\partial v}{\partial t} + v\frac{\partial v}{\partial z} = \eta(E_{\mathrm{c}} + E_{\mathrm{sc}}), \tag{10.6}$$

where the expansion of the left-hand side recognizes that the rate of change of the velocity of the electrons can be expressed as the sum of the rate of change with time at constant position and the rate of change with position as the motion of the electron is followed. The current equation becomes

$$J = \varrho v, \tag{10.7}$$

and Poisson's equation is

$$\frac{\partial E_{\mathrm{sc}}}{\partial z} = \varrho/\varepsilon_0. \tag{10.8}$$

In these equations the vector quantities $\boldsymbol{J}$, $\boldsymbol{v}$ and $\boldsymbol{E}$ are all assumed to possess only z components. The set of equations is non-linear because (10.7) contains a term which is the product of two variables.

The next stage is to linearize the equations by assuming that all the dependent variables can be written in the form

$$a = a_0 + a_1 \exp \mathrm{j}(\omega t - kz), \tag{10.9}$$

where $a_1 \ll a_0$ and ω and k are the same for all variables. The second term assumes that the equations have wave solutions. The small-signal equations are then

$$kJ_1 = \omega\varrho_1 \tag{10.10}$$

$$j(\omega - kv_0)v_1 = \eta(E_c + E_{sc}) \tag{10.11}$$

$$J_1 = \varrho_0 v_1 + \varrho_1 v_0 \tag{10.12}$$

and
$$-jkE_{sc} = \varrho_1/\varepsilon_0. \tag{10.13}$$

For the moment we will assume that the field of the external circuit is zero so that E_{sc} can be eliminated between (10.11) and (10.13) to give

$$(\omega - kv_0)v_1 = \frac{\eta\varrho_1}{k\varepsilon_0}. \tag{10.14}$$

Similarly J_1 can be eliminated between (10.10) and (10.12) to give

$$(\omega - kv_0)\varrho_1 = k\varrho_0 v_1. \tag{10.15}$$

These two equations can be satisfied simultaneously only if

$$(\omega - kv_0)^2 = \eta\varrho_0/\varepsilon_0. \tag{10.16}$$

The terms of this equation have the dimensions of angular frequency squared so we will set

$$\omega_p^2 = \eta\varrho_0/\varepsilon_0. \tag{10.17}$$

The physical significance of this frequency can be understood by considering the case when the electrons have no d.c. component of velocity so that

$$\omega = \omega_p. \tag{10.18}$$

If a stationary cloud of electrons which is in equilibrium is disturbed then it will oscillate with the frequency given by (10.18). Such a cloud of electrons is known as an electron plasma and ω_p is known as the plasma frequency. The possible solutions to (10.16) can therefore be written

$$k_\pm = \frac{\omega \mp \omega_p}{v_0} = k_e \mp k_p. \tag{10.19}$$

It is convenient to display these solutions in the form of a dispersion diagram (see p. 38) as shown in Fig. 10.2. The slope of each line is equal to the d.c. beam velocity v_0. These lines represent two possible wave solutions whose phase velocities are one greater than and the other less than v_0. The waves are compressional charge density waves and are therefore known as fast and slow space-charge waves.

It is often useful to represent these waves by equivalent transmission-line modes. To do this we define an a.c. voltage known as the beam kinetic voltage V_1 by invoking the principle of conservation of energy

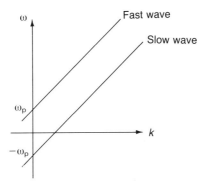

Fig. 10.2 Dispersion diagram of space-charge waves on an electron beam.

$$\tfrac{1}{2}m(v_0 + v_1)^2 = q(V_0 + V_1).\qquad(10.20)$$

For small signals the left-hand side can be expanded to give

$$v_0^2 + 2v_0v_1 = 2\eta(V_0 + V_1),\qquad(10.21)$$

where the second-order term v_1^2 has been neglected. This equation is true at all times and so the a.c. and d.c. terms must balance separately to give

$$V_1 = v_0v_1/\eta\qquad(10.22)$$

and $$V_0 = v_0^2/2\eta.\qquad(10.23)$$

Dividing (10.22) by (10.23) yields

$$\frac{V_1}{V_0} = \frac{v_1}{2v_0}.\qquad(10.24)$$

The power density in the waves can then be calculated using the familiar transmission line equation

$$P = \tfrac{1}{2}V_1I_1^*.\qquad(10.25)$$

From (10.10) and (10.12) the a.c. current is

$$I_1 = \frac{\omega\varrho_0 A}{(\omega - kv_0)}\,v_1,\qquad(10.26)$$

where A is the cross-sectional area of the beam and, making use of (10.16) and (10.24),

$$I_1 = \frac{2\omega\varrho_0 v_0 A}{\pm\omega_p}\left(\frac{v_1}{2v_0}\right)$$

$$= \pm\frac{2\omega I_0}{\omega_p V_0}\left(\frac{V_1}{V_0}\right)\qquad(10.27)$$

so that the power flow in the two waves is

$$P_{\pm} = \frac{1}{2}\frac{|V_1|^2}{Z_{\pm}},\tag{10.28}$$

where the plus and minus signs refer to the fast and slow space-charge waves, respectively. Their characteristic impedances are given by

$$Z_{\pm} = \pm Z_e = \pm 2\frac{\omega_p}{\omega}\frac{V_0}{I_0}.\tag{10.29}$$

Equation (10.28) reveals the surprising fact that the slow space-charge wave carries negative power. The direction of the power flow is certainly positive because, as Fig. 10.2 shows, the slow wave has a positive group velocity. The explanation of this paradox is to be found in equation (10.27) which shows that in the slow wave the a.c. current and velocity are in antiphase. Thus when $v > v_0$, $J < J_0$, and when $v < v_0$, $J > J_0$. It follows that the average kinetic energy of a beam carrying a slow wave is less than the kinetic energy of an unmodulated beam. In order to set up a slow wave it is necessary to remove power from the beam. This unexpected feature of the slow space-charge wave provides the key to understanding all kinds of microwave linear-beam tubes.

An analysis of the propagation of waves along an electron beam confined by an axial magnetic field shows that there are two other modes which can propagate whose propagation constants are given by

$$k_{\pm} = \frac{\omega \mp \omega_c}{v_0},\tag{10.30}$$

where $\omega_c = \eta B$ is known as the electron cyclotron frequency. These waves are known as the fast and slow cyclotron waves. They are associated with motion of the electrons in circular orbits around the direction of the axial magnetic field. A fuller discussion of cyclotron modes is given by Louisell (1960).

10.3 KLYSTRON AMPLIFIERS

The arrangement of a simple klystron amplifier is shown schematically in Fig. 10.3. The electron beam passes along the axes of two re-entrant cylindrical resonant cavities. At the centre of each cavity it crosses a gap where it interacts with the alternating electric field of the cavity. Radio-frequency power is fed into the first cavity setting up fields which modulate the velocities of the electrons. To simplify matters we will assume that the length of the gap is short and that at each end the electron beam passes through a grid of fine wires. These grids have the effect of confining all the gap field to the space between them. In practical klystrons grids cannot be

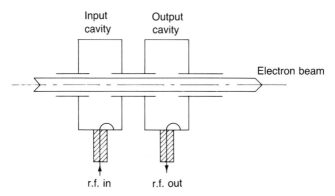

Fig. 10.3 Two-cavity klystron amplifier.

used because they would not be able to withstand the power dissipated in them by the electron beam. It can be shown, however, that gridless gaps can be represented by equivalent gridded gaps for the purposes of analysis.

Figure 10.4 shows a pair of grids d apart with a voltage $V_g\, e^{j\omega t}$ applied across them. To calculate the amplitude of the modulation of the electron velocity we consider an electron which crosses the centre of the gap at the instant when the field is a maximum. Between the grids the electric field is uniform and equal to V_g/d. The equation of motion is then

$$\frac{dv}{dt} = \eta\, \frac{V_g}{d}\, e^{j\omega t} \tag{10.31}$$

so that the amplitude of the velocity modulation produced by the gap is

$$v_1 = \frac{\eta V_g}{d} \int_{t_1}^{t_2} e^{j\omega t}\, dt, \tag{10.32}$$

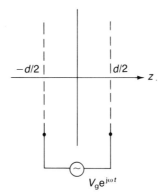

Fig. 10.4 Gridded gap for velocity modulating an electron beam.

where t_1 is the time at which the electron enters the gap and t_2 is the time at which it leaves. If the change in velocity is small compared with the drift velocity v_0 then

$$z = v_0 t \tag{10.33}$$

so that

$$v_1 = \frac{\eta V_g}{v_0 d} \int_{-d/2}^{d/2} e^{jk_e z} \, dz, \tag{10.34}$$

where $k_e = \omega/v_0$. Carrying out the integration we obtain

$$v_1 = \frac{\eta V_g}{v_0} \frac{\sin (k_e d/2)}{k_e d/2}. \tag{10.35}$$

This can be expressed in terms of the beam kinetic voltage by using (10.22) to give

$$V_1 = M V_g, \tag{10.36}$$

where

$$M = \frac{\sin (k_e d/2)}{k_e d/2} \tag{10.37}$$

is known as the gap coupling factor. For an ideal narrow gap $d \to 0$ and $M \to 1$. Physically the coupling factor takes account of the change of the strength of the electric field during the time it takes for an electron to cross the gap. The current is continuous so

$$I_1 = 0. \tag{10.38}$$

Equations (10.36) and (10.38) are the boundary conditions governing the launching of space-charge waves by the gap. Now the general solutions for wave propagation of the beam are

$$V_1 = V_+ \exp j(\omega t - k_+ z) + V_- \exp j(\omega t - k_- z) \tag{10.39}$$

and

$$I_1 = \frac{V_+}{Z_e} \exp j(\omega t - k_+ z) - \frac{V_-}{Z_e} \exp j(\omega t - k_- z), \tag{10.40}$$

where $k_\pm$ and Z_e are given by (10.19) and (10.29). Using the boundary conditions we see that $V_\pm = \frac{1}{2} V_g$. The kinetic voltage and current on the beam are therefore

$$
\begin{aligned}
V_1 &= \tfrac{1}{2} M V_g (e^{jk_p z} + e^{-jk_p z}) \exp j(\omega t - k_e z) \\
&= M V_g \cos k_p z \exp j(\omega t - k_e z),
\end{aligned} \tag{10.41}
$$

where $k_e = \omega/v_0$, $k_p = \omega_p/v_0$, and

$$I_1 = -\frac{1}{2}\frac{MV_g}{Z_e}(e^{jk_pz} - e^{-jk_pz})\exp j(\omega t - k_e z)$$

$$= -j\frac{MV_g}{Z_e}\sin k_p z \exp j(\omega t - k_e z). \tag{10.42}$$

From these two equations it can be seen that as the waves propagate down the beam the amplitudes of the voltage and current modulations change sinusoidally. The wavelength of these standing waves is the plasma wavelength given by

$$\lambda_p = \frac{2\pi}{k_p}. \tag{10.43}$$

Thus, at a distance $\frac{1}{4}\lambda_p$ from the input gap the voltage (i.e. velocity) modulation is zero and the current modulation is a maximum. At this plane the charge density in the beam varies sinusoidally with time and the beam is said to be bunched.

When a bunched beam crosses an interaction gap it induces a current in the gap. The way in which this happens can be understood by considering Fig. 10.5. A bunch of electrons induces positive charges on the grids bounding the gap. As the bunch moves across the gap the induced charges redistribute themselves flowing from one grid to the other through the external load R. The charges shown in the diagram should be regarded as a.c. charges, that is they are charges relative to the d.c. level. Half a cycle later than the situation shown in the diagram the charge in the gap will be less than the d.c. level so it can be thought of as a positive a.c. charge. The induced charges will then be negative and the induced current will flow in the opposite direction.

The flow of induced current through the external load resistor produces a potential difference across the gap and an electric field in the direction shown. This field acts to slow down the electrons in the bunch. The energy removed from the beam is dissipated in the load resistance. This fact enables us to calculate the relationship between the induced current and the

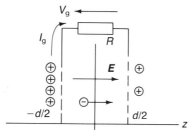

Fig. 10.5 Induction of current by the motion of a bunch of electrons across a gridded gap.

a.c. component of the beam current. At the moment when the centre of the bunch is at the centre of the gap the charge distribution in the gap is given by

$$\varrho(z) = I_1(z)/v_0 \tag{10.44}$$

since the a.c. component of velocity is zero when $\sin k_p z = 1$. From (10.44) we can write

$$\varrho(z) \simeq \frac{|I_1|}{v_0} e^{-jk_e z} \tag{10.45}$$

since $k_p \ll k_e$ so that the sine term in (10.42) is very close to unity. The force on an element of charge within the gap is then

$$F = \int_{-d/2}^{d/2} E_\varrho(z) dz \tag{10.46}$$

so that the rate at which energy is being extracted from the bunch is

$$Fv_0 = \int_{-d/2}^{d/2} E|I_1| e^{-jk_e z} \, dz. \tag{10.47}$$

But this must exactly balance the rate of dissipation of energy in the load given by $I_g V_g$. Thus

$$I_g V_g = \frac{V_g}{d} |I_1| \int_{-d/2}^{d/2} e^{-jk_e z} \, dz \tag{10.48}$$

so that

$$I_g = |I_1| \frac{\sin k_e d/2}{k_e d/2}$$
$$I_g = M|I_1|. \tag{10.49}$$

Thus the gap coupling factor also controls the effectiveness with which the beam drives the gap. It should be noted that the modulated beam behaves as a near-ideal current source.

In the derivation of equations (10.36) and (10.49) we have neglected the space-charge forces. This is justifiable because the space-charge field is normally much smaller than the gap field for both input and output gaps. There is, however, one factor which this analysis does not reveal because it supposes that both space-charge waves are excited with equal amplitudes. Since in that case they carry equal and opposite powers it would appear that no power is needed to modulate the beam and the power gain of the device must be infinite. If the analysis of energy extraction from the beam were repeated taking account of the difference between the space charge waves the result would be

$$I_g = M_+|I_+| + M_-|I_-| \tag{10.50}$$

where

$$M_{\pm} = \frac{\sin k_{\pm}d/2}{k_{\pm}d/2}.$$ (10.51)

It can be shown that these coupling factors are also valid for the input gap. The power required to modulated the beam is therefore the difference between the power carried by the two space-charge waves

$$P_{in} = \frac{1}{2Z_e}\left[\left(\frac{M_+V_g}{2}\right)^2 - \left(\frac{M_-V_g}{2}\right)^2\right]$$

$$= \frac{1}{8}\frac{V_g^2}{Z_e}(M_+^2 - M_-^2).$$ (10.52)

Thus the beam appears to the input gap as an impedance

$$R_b = \frac{4Z_e}{M_+^2 - M_-^2}.$$ (10.53)

The beam loading of the gaps also has a reactive component which only serves to tune the resonances of the cavities slightly.

The complete two-cavity klystron can be modelled by the equivalent circuit shown in Fig. 10.6. The various resistive components are represented by their conductances ($G = 1/R$). Thus G_s is the source conductance, G_L the load conductance and G_b the beam-loading conductance. The conductance G_c represents the impedance of the cavity resonators. If these are assumed to be tuned to the signal frequency then G_c represents the cavity losses. The magnitudes of the conductances are chosen to give the cavity Q factors appropriate to the bandwidth required. The loaded Q of klystron cavities is typically around 100. The input and output cavities are matched to the source and load when

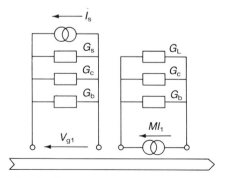

Fig. 10.6 Equivalent circuit of a two-cavity klystron amplifier.

$$G_s = G_c + G_b = G_L. \tag{10.54}$$

The input gap voltage is then

$$V_g = \tfrac{1}{2}I_s/G_s \tag{10.55}$$

and the input power is

$$P_{in} = \tfrac{1}{8}I_s^2/G_s. \tag{10.56}$$

From (10.42) the amplitude of the a.c. beam current at the output gap is

$$|I_1| = MV_g/Z_e \tag{10.57}$$

so that the output power is

$$P_{out} = \tfrac{1}{8}|MI_1|^2/G_s \tag{10.58}$$

whence the power gain is

$$\frac{P_{out}}{P_{in}} = \frac{M^4}{4G_s^2 Z_e^2}. \tag{10.59}$$

For typical values of the parameters the gain would be in the region of 15 to 20 dB.

The gain of a klystron can be increased by adding more cavities at intervals of $\lambda_p/4$ as shown in Fig. 10.7. The loads of the second and third cavities do not dissipate much power. Their function is to adjust the Q of the cavities to give the desired bandwidth. The velocity modulation at the input gap produces current modulation I at the second cavity. The induced current in that cavity produces a gap voltage which, in turn, produces a velocity modulation V. Because the electron beam is a linear system at small signal levels this modulation is added to that already on the beam. The gain between the first and second cavities ensures that this modulation is stronger than the initial modulation. At the third cavity the modulation produced by the first cavity appears as pure velocity modulation whilst that from the second cavity produces current modulation I_3. The process here is exactly like that at the second cavity so that a velocity modulation V_3 is added to the beam. This is the dominant modulation at the output cavity where it produces a current modulation I_4. Thus the gain of a multi-cavity klystron can be computed by a straightforward extension of the method

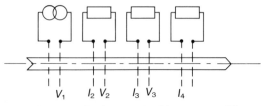

Fig. 10.7 Schematic diagram of a four-cavity klystron amplifier.

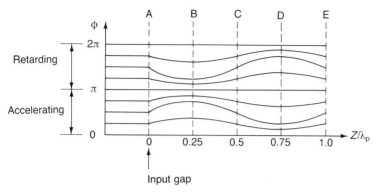

Fig. 10.8 Motion of electrons with velocity modulation relative to the average electron velocity.

described above. The gain at frequencies other than the centre frequency, or when the cavities are not all tuned to the same frequency, can be computed by using matrix algebra to represent the cavities and the drift regions.

The interchange between velocity and current modulation can be represented graphically as shown in Fig. 10.8. This figure shows the phases at which a number of sample electrons cross planes along the length of the beam. The phase is referred to the phase of an electron travelling with the d.c. beam velocity. The slopes of the trajectories show the velocities of the electrons relative to v_0.

At the input gap (A) the trajectories are equally spaced showing no current modulation but their slopes differ because of the velocity modulation produced by the gap field. As the electrons move towards B their trajectories converge. This process is resisted by the repulsive space-charge forces so that at B the trajectories are parallel to the axis. At this plane there is therefore no velocity modulation but a maximum of current modulation. Following the motion forward we see that the space-charge forces cause the trajectories to diverge giving velocity modulation without current modulation at C and so on. From the point of view of an observer travelling with the d.c. beam velocity the electrons are executing oscillations at the plasma frequency about their mean positions. The separation of the planes at which the velocity and current modulations are maximum is therefore equal to $\lambda_p/4$ and is independent of the strength of the modulation.

Clearly it is not possible for this situation to continue at ever higher signal levels. Eventually the process becomes non-linear and the simple theory given above breaks down. This is illustrated by Fig. 10.9. As the signal level increases the plane at which the current modulation is a maximum moves back towards the input gap so the optimum drift length is less than $\lambda_p/4$ (Fig. 10.9(a)). When the signal level is increased still further some of the electron trajectories cross over each other (Fig. 10.9(b)). Once crossing

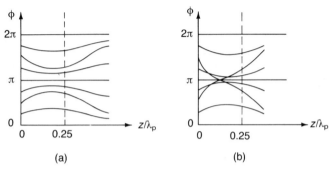

Fig. 10.9 Motion of electrons with velocity modulation showing: (a) shortening of the bunching length, and (b) electron crossovers at high modulation levels.

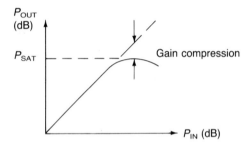

Fig. 10.10 Transfer characteristic of a klystron showing saturation at high drive levels.

over has occurred the effect of the space-charge field is to force the electron still further away from the bunch. This process sets a limit to the maximum bunching which can be achieved and thus to the maximum output power and efficiency of the device. Figure 10.10 shows the typical form of the transfer characteristic of a klystron. At low signal levels the device is a true linear amplifier over a wide dynamic range.

At high signal levels the output saturates as shown. The gain compression at saturation is about 5 dB. For maximum efficiency a klystron must be operated at saturation and this is normal for tubes used for radar transmitters which use an unmodulated carrier. For communications systems such as television transmitters the tube must be operated below saturation in order to avoid non-linear effects.

10.4 SLOW-WAVE STRUCTURES

A klystron works through the interaction between an electron beam and resonant, standing, electromagnetic waves. Travelling-wave tubes (TWTs), on the other hand, employ travelling waves as their name suggests. Because

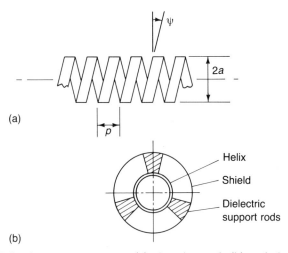

Fig. 10.11 Helix slow-wave structure: (a) elevation and, (b) end elevation.

these waves are non-resonant the interaction can take place over a much wider band of frequencies than in a klystron. In order for an electron beam to interact with a travelling wave it is necessary that the phase velocity of the wave should be approximately the same as the velocity of the electrons.

Waves propagating in a hollow waveguide have phase velocities greater than the velocity of light. They are therefore unsuitable for travelling-wave interactions. To produce a suitable wave we must use a slow-wave structure of which there are two main types.

The first is the helix shown in Fig. 10.11. Modern TWTs generally use a tape helix as shown in Fig. 10.11(a). This helix is mounted within a concentric conducting shield by means of dielectric support rods as shown in Fig. 10.11(b). The result is a rather special kind of coaxial transmission line. At high frequencies the signal follows the turns of the helix travelling at about the velocity of light. The velocity of the wave along the axis is then

$$v_p = c \sin \psi. \tag{10.60}$$

At low frequencies the wave tends to skip from turn to turn of the helix and the phase velocity tends to the velocity of light. Figure 10.12 shows the dispersion diagram for a helical slow-wave structure. The useful bandwidth of such a slow-wave structure is limited by its dispersion at low frequencies and by possible interaction with other modes of the helix for phase shifts greater than about 180° per turn. A bandwidth of an octave is readily attainable. By using dispersion shaping techniques (Webb, 1985) it is possible to make tubes which give useful gain over bandwidths greater than three octaves.

The second main class of slow-wave structure can be thought of as a

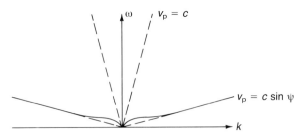

Fig. 10.12 Dispersion diagram of a helix slow-wave structure.

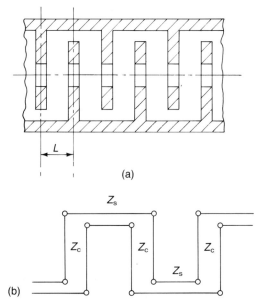

Fig. 10.13 Folded waveguide slow-wave structure: (a) cross-sectional view, and (b) transmission-line equivalent circuit.

folded waveguide as shown in Fig. 10.13(a). The wave travels along the folded guide with a phase velocity greater than the speed of light but its phase velocity in the axial direction is much smaller. This structure can be modelled as a transmission line with alternate sections having different impedances as shown in Fig. 10.13(b). The discontinuities at the junctions between the sections of line produce coupling between the forward and backward waves with stop bands wherever the structure period is an integral number of half wavelengths. The dispersion diagram for a coupled-cavity slow-wave structure of this kind is shown in Fig. 10.14. In this diagram ϕ is the phase shift per section along the folded waveguide. The dispersion

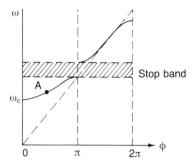

Fig. 10.14 Dispersion diagram of a coupled-cavity slow-wave structure in terms of phase shift along the folded waveguide.

curve is basically the same as that for a hollow waveguide (Fig. 2.16) with propagation at frequencies above the cut-off frequency ω_c. At the edges of the stop band the group velocity is zero. This corresponds to a situation where the effect of reflections at the discontinuities is to produce forward and backward waves of equal amplitude, that is, standing waves. There are two possible standing waves: one with the maximum electric field across the centres of the cavities and the other with the maximum across the centres of the coupling slots as shown in Fig. 10.15. Any general wave can be thought of as a combination of these two normal modes. The two modes have the same wavelength as each other but different frequencies. The relative frequencies depend upon the relative magnitudes of the impedances of the sections of line.

For practical purposes it is the phase shift per cavity along the axis of the

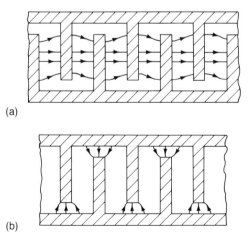

(a)

(b)

Fig. 10.15 Field patterns at the edges of the stop band in a coupled-cavity slow-wave structure.

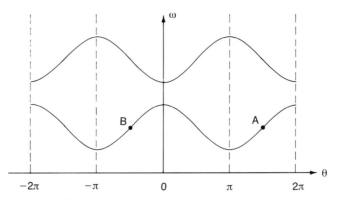

Fig. 10.16 Dispersion diagram of a coupled-cavity slow-wave structure in terms of phase shift per cavity.

slow-wave structure which matters rather than the phase shift along the folded waveguide. A little thought shows that the folding of the waveguide introduces an additional phase shift of 180° per section. Electrons travelling along the axis of the structure experience the interaction field only when they are crossing a cavity. The phase shift perceived by the electrons may therefore be written

$$\theta = \theta_0 + 2n\pi, \tag{10.61}$$

where $n = 0, \pm 1, \pm 2, \pm 3$, etc. and θ_0 is the phase difference between the fields at the centres of adjacent cavities. The dispersion diagram of the slow-wave structure can therefore be plotted in the form shown in Fig. 10.16. The different branches of the dispersion curve are repeated periodically in both directions by virtue of (10.61). The velocity of the electrons is chosen

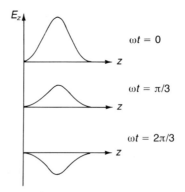

Fig. 10.17 Axial electric field of an interaction gap at different times in the r.f. cycle.

to be synchronous with the circuit wave at the point A where both the group and phase velocities of the circuit wave are positive. The interaction is then similar to that occurring with a helix slow-wave structure.

The significance of the different parts of the curves in Fig. 10.16 becomes clearer if we consider the axial electric field for a particular frequency and phase shift per cavity. Figure 10.17 shows the axial field in one cavity at time intervals corresponding to phase shifts of 60°. If we assume that the phase shift per cavity is also 60° and plot the fields of adjacent cavities the result is a shown in Fig. 10.18. This combination of fields may be expressed as a Fourier series in space and the fundamental component is evidently a wave travelling in the positive z direction as shown by the broken curve in Fig. 10.19(a). At the moment illustrated the magnitude of the field at A is increasing whilst that at B is decreasing both of which are consistent with a fundamental Fourier component travelling to the right.

The next Fourier component is shown in Fig. 10.19(b). From (10.61) this has a phase shift per cavity of $-300°$ ($n = -1$). Consideration of the fields at A and B shows that this wave is travelling in the negative z direction as shown. This is consistent with the negative value of phase shift per cavity computed from (10.61). The wave shown in Fig. 10.19(b) is described as a space-harmonic wave. The complete interaction field can therefore be described in two ways: first as the superposition of a set of standing waves and second as the superposition of a set of travelling space-harmonic waves. For waves propagating in a lossless structure these descriptions are entirely equivalent to each other. In terms of Fig. 10.16 the fundamental waves are in the region $-\pi$ to π whilst the first space-harmonic waves occupy the regions $(-2\pi, -\pi)$ and $(\pi, 2\pi)$. Thus the fundamental forward-wave branch at A in Fig. 10.14 gives rise to a fundamental backward-wave at B in Fig.

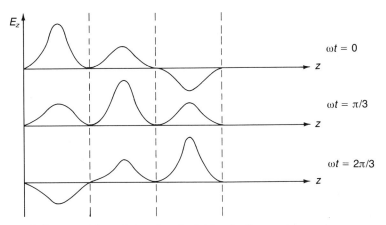

Fig. 10.18 Superposition of the electric fields of adjacent interaction gaps at different times in the r.f. cycle.

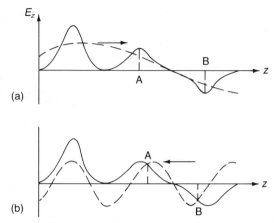

Fig. 10.19 Fourier coefficients of the electric field in a coupled-cavity slow-wave structure: (a) forward-wave fundamental, and (b) backward-wave space harmonic.

10.16 because of the phase reversal introduced by folding the waveguide. The electrons interact with the first forward-wave space-harmonic of that wave at the point A in Fig. 10.16.

10.5 TRAVELLING-WAVE TUBES

Before proceeding to a theoretical analysis of the travelling-wave tube it is useful to obtain a physical understanding of how it works. Figure 10.20(a) shows a uniform stream of electrons which are travelling synchronously

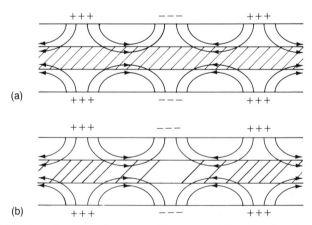

Fig. 10.20 Electron bunching in a travelling wave tube: (a) electric field of the slow-wave structure, and (b) phase relationship between the interaction field and the electron bunches.

with the wave carried by a hypothetical, uniform, slow-wave structure. The electrons are assumed to be constrained by a strong axial magnetic field so that they can only move longitudinally. The fringing field of the slow wave has an axial component of electric field which interacts with the electrons. If the whole system is viewed in a frame of reference which is travelling with the wave then the effect of the interaction is to cause the electrons to become bunched as shown in Fig. 10.20(b). To a stationary observer these current bunches appear as an a.c. current modulation of the electron beam. The electron bunches induce charges on the slow-wave structure which must move along the structure in synchronism with the wave. Moreover, since the bunches are negatively charged, the induced charges have the same polarity as those arising from the original slow wave and add to them to produce growth in the signal level along the length of the tube. The bunching process is limited by the forces of mutual repulsion between the electrons. The principle of conservation of energy requires that the electrons should lose kinetic energy as they transfer power to the wave on the circuit. Eventually this causes the desired synchronous relationship between the electrons and the wave to be lost so that the maximum conversion of energy from d.c. to r.f. is limited.

For the purposes of analysis it is convenient to represent the slow-wave structure by an equivalent transmission line as shown in Fig. 10.21. The inductance and capacitance per unit length of this line can be chosen to make it equivalent to a helix at any given frequency. The propagation of waves on the line coupled to the electron beam are governed by the equations

$$\frac{\partial I}{\partial z} = -C\frac{\partial V}{\partial t} - \frac{\partial i}{\partial z} \tag{10.62}$$

$$\frac{\partial V}{\partial z} = -L\frac{\partial I}{\partial t}. \tag{10.63}$$

There are identical to the usual transmission-line equations (Carter, 1986) with the exception of the last term in (10.62) which reflects the fact that an a.c. current i on the electron beam induces a current on the slow-wave structure and that the total a.c. current on the system must obey Kirchhoff's

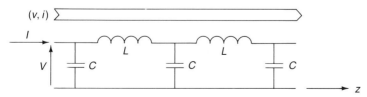

Fig. 10.21 Schematic diagram of a travelling wave tube.

current law. If we assume that waves propagate on the coupled system as $\exp \mathrm{j}(\omega t - kz)$ then we obtain

$$\mathrm{j}kI = \mathrm{j}\omega CV - \mathrm{j}ki \qquad (10.64)$$

$$\mathrm{j}kV = \mathrm{j}\omega LI \qquad (10.65)$$

and, eliminating I, we obtain

$$V = \frac{kk_0 Z_0}{k_0^2 - k^2} i, \qquad (10.66)$$

where

$$k_0 = \omega \sqrt{(LC)} \qquad (10.67)$$

is the natural propagation constant for waves on the line in the absence of the electron beam and

$$Z_0 = \sqrt{(L/C)} \qquad (10.68)$$

is the characteristic impedance of the line. Notice that the inductance and capacitance per unit length have been replaced by quantities which have more meaning at microwave frequencies.

The interaction field is related to the voltage on the line by

$$E_c = -\frac{\partial V}{\partial z} = \mathrm{j}kV \qquad (10.69)$$

so that

$$E_c = \mathrm{j}\frac{k^2 k_0 Z_0}{k_0^2 - k^2} i. \qquad (10.70)$$

A second relationship between the a.c. current on the beam and the a.c. circuit voltage can be derived by considering the electron beam dynamics starting from equations (10.10) to (10.13). Eliminating ϱ_1 from (10.10) and (10.12) and from (10.10) and (10.13) yields

$$(\omega - kv_0)J_1 = \omega \varrho_0 v_1 \qquad (10.71)$$

and

$$E_{\mathrm{sc}} = \mathrm{j}\frac{J_1}{\omega \varepsilon_0}. \qquad (10.72)$$

We also have

$$\mathrm{j}(\omega - kv_0)v_1 = \eta(E_c + E_{\mathrm{sc}}). \qquad (10.11)$$

Eliminating the space-charge field and the a.c. velocity from (10.11), (10.71) and (10.72) gives

$$J_1 = \frac{\mathrm{j}\omega v_0}{\omega_{\mathrm{p}}^2 - (\omega - kv_0)^2}\frac{J_0 E_c}{2V_0} \qquad (10.73)$$

and integrating across the cross sectional area of the electron beam gives

$$i = \frac{j\omega v_0}{\omega_p^2 - (\omega - kv_0)^2} \frac{I_0 E_c}{2V_0} \tag{10.74}$$

in which the d.c. and a.c. currents replace the corresponding current densities.

For self consistency equations (10.66) and (10.74) must be satisfied simultaneously so that the possible values of the propagation constant k are given by

$$[k_0^2 - k^2][k_p^2 - (k_e - k)^2] = -k_e k_0 k^2 \frac{I_0 Z_0}{2V_0}. \tag{10.75}$$

This equation is known as the determinantal equation of the system. If the coupling term on the right-hand side of the equation is set to zero the roots are

$$k = \pm k_0 \tag{10.76}$$

(the forward and backward waves on the transmission line) and

$$k = k_e \pm k_p \tag{10.77}$$

(the fast and slow space-charge waves on the electron beam). These solutions are shown in the dispersion diagram in Fig. 10.22. The interaction which produces gain is that at A between the forward circuit wave and the slow space-charge wave. We have already remarked that the slow space-charge wave is excited by removing energy from the electron beam. It is therefore possible for energy to be transferred from this wave to the wave on the slow-wave structure in such a way that both grow with distance.

The full solution of the coupled equation (10.75) requires the use of a computer but a useful approximation can be derived by assuming that only coupling between the slow space-charge wave and the forward circuit wave

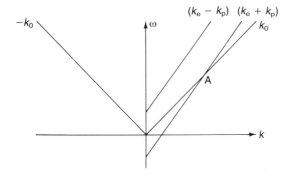

Fig. 10.22 Dispersion diagram of the uncoupled slow-wave structure and electron beam modes in a helix travelling-wave tube.

is important. First (10.75) is rewritten with the left-hand side expressed as a product of four factors and the right-hand side rewritten in terms of the beam characteristic impedance Z_e (10.29) to give

$$(k_0 - k)(k_0 + k)(k_p - k_e + k)(k_p + k_e - k) = -k_e k_0 k k_p \frac{Z_0}{Z_e}. \quad (10.78)$$

This can be rearranged to keep only the two roots in which we are interested on the left-hand side

$$(k - k_0)(k - k_p - k_e) = -\left(\frac{kk_0}{k + k_0}\right)\left(\frac{k_e k_p}{k - k_e + k_p}\right)\frac{Z_0}{Z_e}. \quad (10.79)$$

Now we know that the roots of this equation must lie close to the roots of the left-hand side, namely $k = k_0$ and $k = k_e + k_p$. We therefore approximate the two brackets on the right-hand side of the equation by substituting the two roots into them in the same order. The result is

$$(k - k_0)(k - k_p - k_e) = -\frac{k_0 k_e}{4}\frac{Z_0}{Z_e}. \quad (10.80)$$

This equation can be solved by the usual method to give

$$k = \tfrac{1}{2}(k_0 + k_p + k_e)$$
$$\pm \tfrac{1}{2}[(k_0 - (k_p + k_e))^2 - k_0 k_e Z_0/Z_e]^{\frac{1}{2}}. \quad (10.81)$$

The solution reveals a number of important things about the travelling wave tube interaction.

If the term under the square root is negative k has a pair of complex conjugate roots. These correspond to a pair of waves one of which grows exponentially with distance whilst the other decays. The real part of both these roots is given by the first term which is just the mean of the two uncoupled roots.

At synchronism $k_0 = k_e + k_p$ and (10.81) becomes

$$k = k_0\left[1 \pm \frac{1}{2}j\sqrt{\left(\frac{Z_0}{Z_e}\right)}\right]. \quad (10.82)$$

The imaginary part is then a maximum so the maximum gain per wavelength is given by

$$\text{Gain} = 20 \log_{10}\left[\exp \pi\sqrt{\left(\frac{Z_0}{Z_e}\right)}\right]$$
$$= 27.3\sqrt{\left(\frac{Z_0}{Z_e}\right)} \text{ dB.} \quad (10.83)$$

Typically Z_e is 50 to 100 times Z_0 giving a gain of around 3 to 4 dB per wavelength. It is not surprising that high gain is obtained with high values of Z_0, corresponding to a high interaction field for a given power flow, and low Z_e, corresponding to a high-current electron beam.

The bandwidth of the interaction can be deduced by considering the points at which the imaginary part of (10.80) becomes zero, that is

$$\frac{k_0 - (k_p + k_e)}{\sqrt{(k_0 k_e)}} = \sqrt{\left(\frac{Z_0}{Z_e}\right)}. \qquad (10.84)$$

The left-hand side of this expression is the normalized difference between the propagation constants of the uncoupled waves. This increases with the square root of the ratio of the circuit impedance to the beam impedance. Figure 10.22 shows that for a given value of this ratio the bandwidth increases as the angle between the two intersecting lines decreases, that is, as synchronism is maintained over a broader band of frequencies. In practice these requirements tend to conflict as broad-band slow-wave structures have low impedances.

The information deduced in the preceding paragraphs enables us to sketch the form of the dispersion diagram for the coupled system. This takes the form shown in Fig. 10.23 with complex conjugate roots and gain in the frequency range $\omega_1 < \omega < \omega_2$.

The theory of the coupled-cavity travelling-wave tube takes a slightly different form because the interaction between the beam and the wave is lumped rather than continuous. Figure 10.24 shows the coupled-mode diagram for a coupled-cavity TWT. The mid-band gain of such a tube can be estimated fairly accurately from (10.83) if the impedance of the correct space harmonic is substituted for Z_0. The analogy with the continuous interaction breaks down at low values of phase shift per cavity when the excitation of a backward wave becomes important.

Helix TWTs typically have working bandwidths in the range from one to

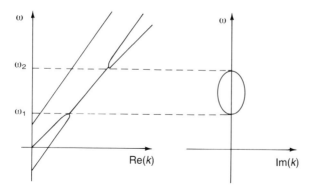

Fig. 10.23 Dispersion diagram of a travelling-wave tube showing the presence of a complex conjugate pair of roots produced by the interaction between the slow space-charge wave and the forward wave on the helix.

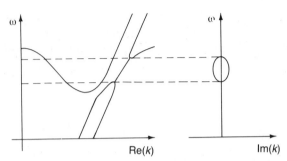

Fig. 10.24 Dispersion diagram for a coupled-cavity travelling-wave tube in which the slow space-charge wave interacts with the first space harmonic of the wave on the slow-wave structure.

three octaves. At 10 GHz mean powers of a few hundred watts and peak, pulsed, powers of a few kilowatts can be obtained. Coupled-cavity TWTs operate over narrower bands (around 15%) because of the greater dispersion of coupled-cavity slow-wave structures. However they have much better heat dissipation capabilities than helices so that a mean power of 10 kW and a peak power of several hundred kilowatts, or more, at 10 GHz are readily achieved. Further information about travelling-wave tubes can be found in Gilmour (1986).

10.6 CROSSED-FIELD TUBES

The linear-beam tubes (klystrons and TWTs) described so far all rely for their operation on a linear electron beam which is confined by some arrangement of axial magnetic field. This is not the only way to get electrons to flow in a smooth controlled manner. A possible alternative is to use crossed electric and magnetic fields as shown in Fig. 10.25(a) with the electric field

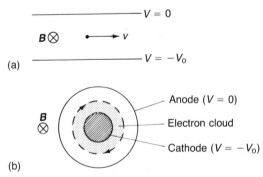

Fig. 10.25 Crossed-field electron flow: (a) linear geometry, and (b) cylindrical geometry.

produced by a pair of plane parallel electrodes. Here the electric and magnetic forces balance and the electron moves in a straight line parallel to the electrodes. In practice it is convenient to make the electrodes concentric cylinders so that the electrons move in regular circular orbits. The addition of space charge does not materially affect the electron flow so it is possible to produce a rotating electron cloud as shown in Fig. 10.25(b). Provided that the magnetic field is strong enough the electrons do not reach the anode and the diode is cut off. The electrons are usually emitted from the whole of the cathode surface through a combination of thermionic and secondary emission.

By arranging that the surface of the anode is a slow-wave structure it is possible to envisage interactions with the electrons analagous to those taking place in a TWT. The mathematical analysis of this kind of device is difficult and they normally operate in the non-linear regime anyway so we will concentrate on the physical principles of operation.

The commonest device of this kind is the magnetron oscillator in which the slow-wave structure is made in the form of a closed circle as shown in Fig. 10.26. The anode and cathode together form a two-conductor transmission line which can propagate TEM waves at the velocity of light and at frequencies down to d.c. The forward and backward waves are coupled by the discontinuities in the structure and are most strongly coupled at a frequency close to that at which the spaces between the vanes are resonant. The result, as in coupled-cavity TWT structures, is to produce a pattern of pass and stop bands as shown in Fig. 10.27. Because the slow-wave structure is closed upon itself the possible frequencies of oscillation are limited to those for which there is a whole number of wavelengths around the perimeter of the device. The possible resonances in the lowest mode are shown in Fig. 10.27 for an eight-vane anode like that in Fig. 10.26. The flat top of the dispersion curve means that most of the resonances are grouped within a very narrow frequency band.

Of all the possible resonances only that for which the phase shift per

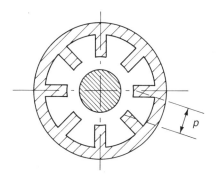

Fig. 10.26 Cross section of a magnetron anode.

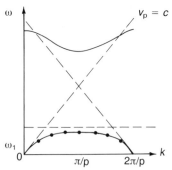

Fig. 10.27 Dispersion diagram of an eight-cavity magnetron anode showing the resonant frequencies.

cavity is 180° has a field pattern which is locked to the orientation of the anode vanes. This 'π mode' is therefore the one chosen for the interaction. When the magnetron is turned on the electrons may interact with other modes besides the π mode particularly the next nearest modes on either side. The interaction starts from the thermal noise in the device and the different modes compete until eventually one becomes dominant. An important part of magnetron design is ensuring that the correct mode is always excited to the exclusion of others. This is achieved by careful design of the anode to get the greatest possible separation between the π mode and the unwanted modes and by tailoring the shape of the voltage pulse applied between cathode and anode to assist anode the growth of the π mode.

In common with other oscillators the magnetron can be locked to a particular phase and frequency by the injection of a signal within the anode resonance Q curve at a level about 10 dB below the output power of the device. A next stage in this train of thought is to make a break in the slow-wave structure, so that it is no longer resonant, and to couple the ends to external waveguides as shown in Fig. 10.28. This device can act as an amplifier employing either a forward-wave or backward-wave interaction depending upon the type of slow-wave structure employed. It does not show the kind of linear, small-signal, amplification found in linear beam tubes and is probably best thought of as a special kind of locked oscillator with a wider range of operating frequencies than the magnetron. The gain is usually only 10 to 20 dB in contrast to the 40 to 60 dB commonly achieved by linear-beam tubes.

Crossed-field devices possess a number of advantages over linear-beam tubes. First, they are much lighter and more compact for a given power output and frequency. Second they can operate at lower impedances (low voltage, high current) and, third, they are intrinsically more efficient converters of d.c. input to r.f. output power. The last property arises because

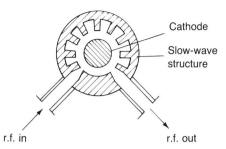

Fig. 10.28 Arrangement of the slow-wave structure of a cross-field amplifier.

the electrons move gradually closer to the anode as they interact with the wave on the structure. Their kinetic energy is thus replenished from the d.c. electric field as fast as it is converted to r.f. power. The electrons therefore remain in synchronism with the wave throughout the interaction. Magnetrons with conversion efficiencies of over 80% have been made for applications such as industrial heating where efficiency is a prime consideration. For further information on crossed-field devices see Gilmour (1986).

10.7 FAST-WAVE DEVICES

All the devices described so far in this chapter depend for their operation upon slow-wave structures or resonant cavities. These features scale linearly with wavelength and become increasingly difficult and expensive to make as the required frequency and power increase. The upper frequency limit for conventional slow-wave devices is around 100 GHz. Alternative approaches which do not require the use of delicate microwave structures have been the subject of much research in recent years because of their potential for generating large amounts of power at sub-millimetre wavelengths.

In a coupled-cavity TWT it is possible to work at beam velocities well below the speed of light by making use of the space harmonics of the wave on the slow-wave structure. This suggests that useful interaction could be obtained by using an electron beam which varies periodically in some fashion so that the interaction is between a fast electromagnetic wave in a waveguide and a space harmonic of the electron beam modes.

Figure 10.29 shows one possible way of achieving this result. A thin high-velocity electron beam confined by an axial magnetic field is directed down the centre of a rectangular waveguide. The beam is deflected alternately to the right and left of the mid-plane by the periodic transverse magnetic field of a 'wiggler'. Each time the direction of transverse motion of an electron is reversed it experiences an acceleration and therefore emits electromagnetic radiation over a broad band of frequencies. If the ends of the waveguide

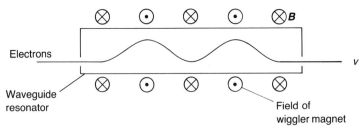

Fig. 10.29 Schematic diagram of a free-electron laser.

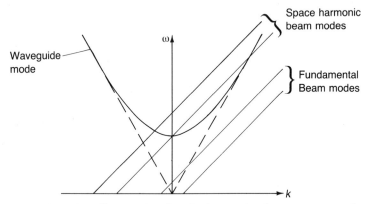

Fig. 10.30 Dispersion diagram showing the interaction between a space-harmonic of the waves on a periodic electron beam and a fast waveguide mode.

are closed so that it becomes resonant then the radiation from the electrons is stimulated by the standing wave in the resonator. The resemblance between this situation and that in a laser has led to the description of this device as a 'free electron laser' (FEL). An alternative view is that the interaction is represented by the dispersion diagram shown in Fig. 10.30 which shows the synchronism between the fast electromagnetic wave in the waveguide and a space harmonic of the slow cyclotron mode on the electron beam. FELs have, so far, only been constructed as experimental devices making use of the intense, high-energy, beams generated by electron accelerators. In one experiment pulsed powers of up to 1 GW were obtained at 35 GHz. In another it has been demonstrated that electromagnetic powers at optical wavelengths can be produced.

An alternative arrangement, shown in Fig. 10.31, has a hollow relativistic electron beam directed along a circular waveguide. The electrons interact with the field of the TE_{01} (circular electric) mode of the waveguide at the cyclotron frequency. They are arranged at the radius where the field strength of that mode is a maximum. Within the beam individual electrons describe orbits as shown in Fig. 10.32. The radius of an orbit is given by

$$r = v_\perp/\omega_c \qquad (10.85)$$

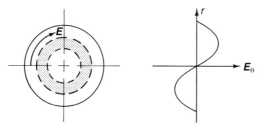

Fig. 10.31 The arrangement of a hollow electron beam in a cylindrical waveguide used in gyrotrons.

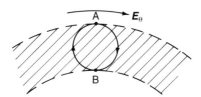

Fig. 10.32 Gyrotron interaction between the tangential component of the r.f. field and an electron moving in a small orbit within the electron beam.

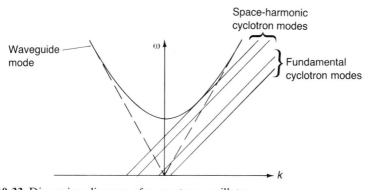

Fig. 10.33 Dispersion diagram of a gyrotron oscillator.

where v_1 is the tangential velocity of the electron and the relativistic cyclotron frequency is given by

$$\omega_c = \frac{\eta B_0}{\sqrt{(1 - v^2/c^2)}}. \tag{10.86}$$

An electron at the point A is accelerated by the field so its cyclotron frequency increases. Conversely an electron at B is retarded so that its cylotron frequency decreases. The effect of this is that the electrons are bunched by the field and can give up energy to it. This is an interaction between the fast cyclotron mode on the beam and the fast electromagnetic wave in the waveguide as shown in Fig. 10.33. Devices which work on this principle are known as cyclotron resonance masers (CRMs) or gyrotrons. Gyrotrons have been constructed which provide 100 kW of pulsed power at 200 GHz. At lower frequencies pulsed powers in excess of a megawatt and mean powers of 75 kW has been achieved. Further information on fast-wave devices can be found in Granatstein and Alexeff (1987).

10.8 ELECTRON ACCELERATORS

The devices described in the earlier parts of this chapter are all concerned with the conversion of the energy of moving electrons into radiofrequency power. It is, however, possible to reverse the processes in order to produce beams of high-energy electrons or other charged particles.

The linear accelerator works rather like a coupled-cavity travelling-wave tube. The difference is that the phase change per cavity is chosen to accelerate the bunches of electrons instead of extracting energy from them. The general arrangement of a travelling-wave linear accelerator is shown in Fig. 10.34. Electrons from an electron gun are injected into the first cavity of a chain of coupled cavities. The cavities are generally coupled together by the axial hole through which the electrons pass. Radiofrequency power is fed in to the cavity chain from a high-power source such as a magnetron to provide a very strong axial r.f. electric field in the cavity. Around half of the injected electrons are captured by the field and formed into bunches which are then accelerated by the fields of subsequent cavities. The cavities are made progressively longer through the first few cells to maintain synchronism with the accelerating bunches. Because of the very high accelerating fields used the velocity of the electrons very quickly approaches the velocity of light. The kinetic energy of the electrons is then given by

$$E = m_0 c^2 \left[\frac{1}{\sqrt{(1 - v^2/c^2)}} - 1 \right], \qquad (10.87)$$

where m_0 is the rest mass of the electron and c is the velocity of light (Rosser, 1964). As the electron velocity approaches the velocity of light the

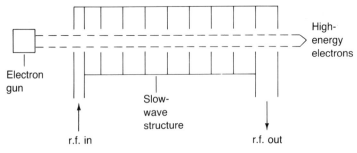

Fig. 10.34 Schematic diagram of a microwave linear accelerator.

kinetic energy becomes very large. Thus, after the first few cells, the effect of the accelerating field is to increase the energy of the electrons without appreciably increasing their velocities. The energies of the electrons are usually expressed in terms of electron volts (the energy gained by an electron in moving through a potential difference of one volt). The electron energies produced by linear accelerators vary from a few MeV for accelerators used for radiotherapy to the 20 GeV of the two-mile accelerator at Stanford in California.

When very high-energy electrons are required from a more compact source a synchrotron may be used. This machine has the general arrangement shown in Fig. 10.35. Electron bunches are formed by a linear accelerator and injected into a beam tube which forms a closed loop. Straight sections of the tube are separated by bending magnets whose strength can be adjusted to keep the bunches moving along the centre of the beam tube as their energy increases. One or more microwave cavities very like those used in klystrons are inserted into the beam tube. The synchrotron is designed in such a way that the bunches experience an accelerating field each time they cross a cavity. Because the electron velocity is very close to the speed of light the time taken for a bunch to go once around the ring is virtually constant. A synchrotron may be used as a source of high-energy electrons (typically a few GeV) or as a source of intense electromagnetic radiation. This radiation is produced by the radial acceleration of the electrons in the bending magnets or by putting a wiggler magnet in one of the straight sections. Synchrotrons and linear accelerators have both been used

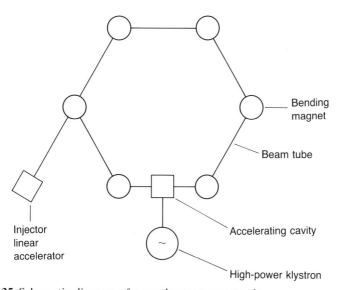

Fig. 10.35 Schematic diagram of a synchrotron storage ring.

to provide the high-energy electrons required for experiments with free-electron lasers.

10.9 CONCLUSION

In this chapter we have considered some of the interactions which take place between free electrons and electromagnetic waves. These interactions are fundamental to the operation of high-power microwave sources and to particle accelerators. Closely related phenomena occur in ionized gasses including the plasmas used in experiments on nuclear fusion.

Streams of electrons carry space-charge and cyclotron waves and these can interact with external r.f. fields to produce amplification. The beating between the fast and slow space-charge waves is used in the klystron. The interaction between the slow space-charge wave and the wave on a slow-wave structure is fundamental to the operation of the travelling-wave tube and the magnetron oscillator.

Newer devices such as the gyratron and the free-electron laser employ interactions between electrons and fast electromagnetic waves in wave-guides. These devices hold the promise of generation of very large amounts of power at wavelengths ranging from millimetres down to optical wavelengths.

The interaction between charged particles and radiofrequency electric fields is employed in linear accelerators, synchrotrons and other particle accelerators to produce particle energies ranging from MeV to GeV.

EXERCISES

10.1 Calculate the velocity, current density and plasma frequency for electron beams having the following parameters:

1. $V = 5\,\text{kV}$, $I = 20\,\text{mA}$, diameter $= 1\,\text{mm}$,
2. $V = 10\,\text{kV}$, $I = 0.5\,\text{A}$, diameter $= 5\,\text{mm}$,
3. $V = 60\,\text{kV}$, $I = 12\,\text{A}$, diameter $= 8\,\text{mm}$.

(Note: at high voltages the beam velocity must be calculated using the relativistically correct equation (10.87) in place of (10.23).)

10.2 Calculate the plasma wavelength and the electronic wavelength and characteristic impedance at 3 GHz, 10 GHz and 60 GHz for each of the beams defined in Question 10.1.

10.3 Calculate the d.c. power and the fast- and slow-wave powers at 10 GHz for each of the beams in Question 10.1 if the beam kinetic voltage is 10% of the d.c. voltage in each case.

10.4 Calculate the gain at 10 GHz and the bandwidth of a two-cavity klystron amplifier whose electron beam has the parameters given in Question 10.1 part 1. The cavities have unloaded Q factors of 500 and R/Q of 800 Ω and the gap transit angles $(k_e d)$ are $\pi/2$.

10.5 The electron beam defined in Question 10.1 part 1 is used in a helix travelling-wave tube with a centre frequency of 10 GHz. Calculate the pitch angle of the helix and the gain of a section of tube 30 mm long if the interaction impedance of the helix is 500 Ω.

11 | Microwave measurements

11.1 INTRODUCTION

The quantities which are measured at microwave frequencies are essentially the same as those measured at lower frequencies. But, because the wavelengths of the signals are comparable with the dimensions of the equipment, it is not possible to use the same techniques. In this chapter we shall examine how frequency, signal level, impedance, attenuation and other quantities are measured at microwave frequencies.

All measurements contain sources of error and it is important in any particular case to know what these are and to have an estimate of the magnitude. Strictly speaking no measurement is of any value unless an estimate of its accuracy can be given. We shall therefore pay attention to the errors which occur in microwave measuring systems in the discussion which follows.

11.2 MEASUREMENT OF FREQUENCY

One simple way of measuring frequency is to measure the wavelength of a standing wave on an airspaced coaxial line. This wavelength is half the free-space wavelength so the frequency can be calculated. Figure 11.1 shows the arrangement of a slotted coaxial line. The strength of the electric field on the line is sampled by a wire probe which protrudes a short way into the space between the conductors. The signal picked up is passed via a detector diode and a sensitive amplifier to a meter. The probe must draw some current in order for a measurement to be possible. This, therefore, limits the accuracy of the measurement. As the probe is moved along the line maxima and minima of the standing wave are detected. For a perfect standing wave the minima are zeroes and their positions can therefore be determined with considerable accuracy. In practice there is some uncertainty about the position of a minimum beause the signal detected falls below the noise level of the detection system. This error can be reduced by measuring the positions of as many minima as possible. In that way several

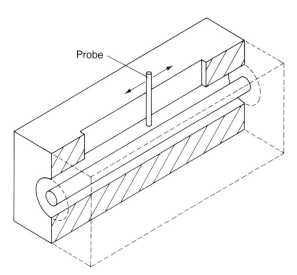

Fig. 11.1 Sectioned view of a slotted section of coaxial line.

different values for the wavelength can be obtained and the average taken to reduce the standard deviation. (Topping, 1962). This approach has the advantage of directness but the accuracy which can be obtained is low (perhaps 0.1% at best) and the measurements are time consuming.

For most microwave laboratory measurements it is much better to have a direct reading of frequency. Originally this was achieved by using a calibrated resonant cavity. By careful design of the cavity the Q factor could be kept high to give a sharp response. The tuning mechanism could also be made to give a direct reading of frequency. Cavity resonance wavemeters, as these devices are called, are still sometimes encountered but they have been superceded by microwave frequency counters. The accuracy of a cavity resonance wavemeter is typically 0.1%.

Microwave frequencies are too high for it to be possible to use the direct counting technique employed at lower frequencies. The way around this is to mix the signal to be measured with that from a crystal controlled local oscillator as shown in Fig. 11.2. If the local oscillator waveform is rich in harmonics then the output from the mixer will be a set of frequencies given by

$$f_i = f_x - nf_1, \tag{11.1}$$

where n is the order of the harmonic and it is assumed that $f_x > nf_i$. The mixer output is fed through a bandpass filter which selects just one frequency out of the set generated. This frequency can be chosen to be low enough for it to be measured with a conventional counter. Since f_1 is known it is possible to compute the source frequency f_x if n can be determined. To

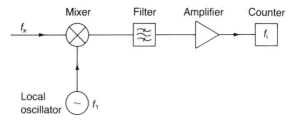

Fig. 11.2 Block diagram of a microwave frequency counter.

do this a second measurement is taken with the local oscillator frequency reduced to f_2 where the offset $(f_1 - f_2)$ is known. The unknown frequency is then given by

$$f_x = nf_1 + f_{i1} \tag{11.2}$$

and
$$f_x = nf_2 + f_{i2} \tag{11.3}$$

where it is assumed that the frequency offset is small enough so that the same harmonic is responsible for the output measured. Eliminating f_x from these two equations gives

$$n = \frac{f_{i2} - f_{i1}}{f_1 - f_2} \tag{11.4}$$

so that n can be computed. The unknown frequency can then be found by substitution back into (11.2).

In practice it is necessary for the method to be a little more complicated to take account of the possibility that one or both of the harmonic frequencies may lie above the unknown. It is also necessary to take steps to ensure that the measurement is accurate even if the incoming signal is frequency modulated.

11.3 MEASUREMENT OF POWER

When a simple detection of a microwave signal is required it is usual to employ a semiconductor diode of the kind shown in Fig. 9.14. At frequencies above 1 GHz it becomes difficult to match the diode satisfactorily because its impedance varies with power level and alternative techniques based on converting the microwave power into heat are used.

At low power levels (a few milliwatts) the detecting element is either a thermistor or a bolometer. A thermistor is manufactured from a mixture of semiconducting oxides and has a negative temperature coefficient of resistance. A bolometer is a thin film resistor deposited on an insulating substrate. Bolometers have response times of less than a millisecond but are very easily damaged by being exposed to too much power.

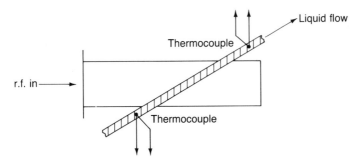

Fig. 11.3 A liquid-flow calorimeter for measuring microwave power.

Thermistors are more rugged but have response times of up to a second. In either case the resistance of the sensing element varies with ambient temperature as well as with the microwave power absorbed. A power meter head therefore normally incorporates two matched thermistors or bolometers which are connected to two arms of a Wheatstone bridge. Only one of the devices is exposed to microwave power. The result is that the balance of the bridge is unaffected by changes in ambient temperature. The bridge is balanced automatically and the output displayed directly in milliwatts on a meter.

At higher power levels (a few watts) the power meter head must be protected from the full power by a calibrated attenuator which is capable of dissipating the full power. An alternative technique is to use a directional coupler to sample the power.

Direct measurement of high power levels is carried out by using a continuous-flow calorimeter as shown in Fig. 11.3. The input power (normally in a waveguide) is absorbed by liquid flowing in a dielectric tube. The tube crosses the guide at an oblique angle to ensure a good match. Very often water flowing in a glass tube is used. The temperature rise in the liquid is measured by a pair of thermocouples. The device is calibrated for a particular flow rate and the flow rate carefully controlled. Alternatively an electric heating element is used as a calibrating heat source.

11.4 MEASUREMENT OF GAIN AND LOSS

In many microwave systems it is necessary to know the gain or loss of each component in order to compute the system performance. These quantities are commonly measured by comparison with standard attenuators. The two possible configurations shown in Fig. 11.4 are r.f. and d.c. substitution. In both cases there is a signal source, a standard attenuator, a detector and some kind of signal level indicator. In r.f. substitution (Fig. 11.4(a)) the attenuator would be a rotary-vane attenuator in a waveguide or a switched

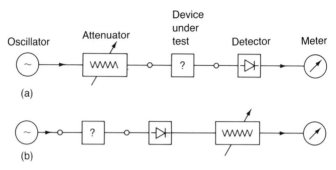

Fig. 11.4 Block diagrams of systems for measuring microwave attenuation: (a) r.f. substitution, and (b) d.c. substitution.

attenuator in a coaxial line. In d.c. substitution (Fig. 11.4(b)) the attenuator could be made in the form of a switched network of precision resistors. The indicator could be a meter or an oscilloscope. The procedure in either case is to set the signal level to a convenient value with the device under test (DUT) in position. It is then removed and the attenuator adjusted to bring the signal back to the same level. This method avoids errors caused by non-linearity in the detector.

In general the gain or loss measured is made up of two components namely that caused by the gain or attenuation inherent in the device and that caused by reflection at mismatches. As the device under test can never be perfectly matched some of the input signal is reflected back towards the source at both the input and the output terminals. Unless the source is very well matched to the connecting transmission lines there will be multiple reflections of the signal producing errors which vary with frequency. A common practice is to put a 10 dB attenuator (a 'pad') between the source and the system to reduce the possibility of multiple reflections as far as possible.

Frequently the measurement is to be made over a band of frequencies. The signal source would then be a sweep oscillator set to sweep repeatedly over the band required and the output could be fed to an $x-y$ plotter to provide a permanent record. A simple r.f. substitution system might use a power meter or a VSWR meter as a detector as shown in Fig. 11.5. Because the output of the oscillator and the sensitivity of the detector vary with frequency it is necessary to produce a set of calibration lines with the attenuator. The performance of the device under test can then be deduced by interpolation between them as shown in Fig. 11.5. Commonly the oscillator is levelled by an external or internal feedback loop to reduce the variation of its output power with frequency.

Better plots of the gain or loss against frequency can be produced if a scalar network analyser system is used. Figure 11.6 shows the general

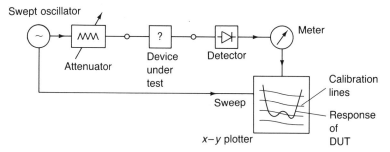

Fig. 11.5 Block diagram of equipment for making swept-frequency insertion-loss measurements.

arrangement. The signal from the sweep oscillator is sampled by high-directivity directional couplers before and after passing through the device under test. The signals in the coupler side-arms are detected and passed to the scalar analyser which is able to display the two signal levels and their ratio in dB against frequency. The output from the scalar analyser can be fed to an $x–y$ plotter to provide a permanent record of the performance of the device under test. The signal-to-noise ratio of the system is enhanced by square-wave modulation of the signal and the use of a tuned amplifier in the scalar analyser.

This arrangement removes errors produced by variations in the output of the oscillator by taking the ratio of the signal levels. It is still liable to errors from a number of sources including the finite directivity of the couplers and any differences in the frequency responses of the couplers and the detectors. Systematic errors which are independent of frequency can be eliminated by removing the device under test and setting the zero level on the analyser. Some systems incorporate a storage normalizer which is able to store the characteristics of the system in the absence of the device under test and correct for them when the result of the measurement is displayed. It is tempting to regard the results produced by such a system as being free from errors though this can never be the case. If the device under test has a high reflection coefficient, for example, then the measurements will be appreciably affected by multiple reflections between it and the source (which can never be a perfect match).

Example

In the system shown in Fig. 11.6 the reflection coefficients of the device under test and the source are 0.3 and 0.05 respectively and the couplers have 20 dB coupling and 40 dB directivity. Investigate the possible errors in the measurement of the insertion loss if all other components can regarded as perfect.

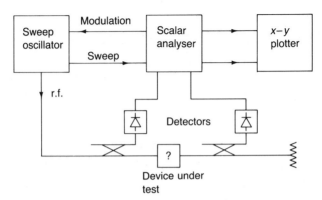

Fig. 11.6 Block diagram of a scalar analyser system for making insertion-loss measurements.

Solution

Multiple reflection of the incident signal between the device under test and the source is illustrated in Fig. 11.7(a). The incident signal level at the device under test is therefore

$$V_1 = \frac{V_0}{1 \pm \varrho_0 \varrho_1}, \qquad (11.5)$$

where the positive and negative signs represent the extreme cases of the phase of the reflected signal relative to V_1. V_1 can therefore vary by ± 0.015 (0.13 dB) about its nominal level.

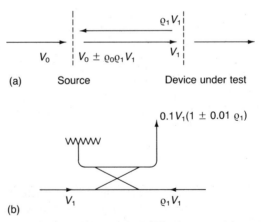

Fig. 11.7 Sources of error in microwave measurements: (a) multiple reflections between the source and the device under test, and (b) finite directivity of directional couplers.

Since the couplers have 40 dB directivity the signal level in the side arm produced by the backward wave is 0.01 of that of a forward wave of the same amplitude. Thus the signal detected is $0.1(1 \pm 0.003)V_1$, as shown in Fig. 11.7(b), giving a possible variation of ± 0.026 dB depending upon the relative phases of the signals. The variation is likely to be below the noise level of the equipment.

Conservation of power at the input port of the device under test requires that the actual signal input is $(1 - 0.09)V_1$. If we suppose that the insertion loss (including any effects of internal reflections) is nominally 20 dB then the signal detected by the second detector is $0.01(1 - 0.09)V_1$. The scalar analyser compares the signal levels at the two detectors. The result will be in error to the extent that the analyser does not form an exact ratio of the signals.

In this example the errors calculated would be likely to lie below the noise level of the system so that they would not be detectable. In other cases this might not be so and it is necessary to be aware of the ways in which the result of a measurement might be in error. For more complicated microwave systems the method of signal flow graphs (Seely, 1972) is used to analyse the errors.

11.5 MEASUREMENT OF RETURN LOSS

A simple modification to the system shown in Fig. 11.6 allows us to measure the return loss of a component directly (see Fig. 11.8). This time the two directional couplers are set to measure the incident and reflected power in the transmission line connected to the input port of the device under test. The arrangement, known as a reflectometer, is widely used for the adjustment of the matches of devices during manufacture. The errors involved in this measuring system can be estimated in the way illustrated in the example above. Notice that errors caused by the directivity of the coupler detecting

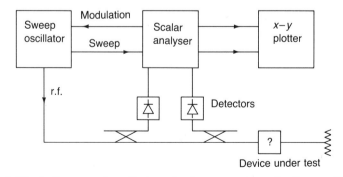

Fig. 11.8 Block diagram of a reflectometer for measuring reflection coefficients.

the reflected power increase as the match of the device under test is improved. Note also that if the transmission loss of the device is caused by internal reflections rather than by dissipation of power then the result will be affected by match of the matched load used to terminate the output line.

11.6 MEASUREMENT OF IMPEDANCE

Although a reflectometer can be a very useful tool for checking and adjusting the match of a device it suffers from the disadvantage that it cannot provide any information about the phase of the reflection. If the requirement is to match a completely unknown impedance then phase information is often necessary.

A basic technique which is still useful though rather tedious is to use a slotted line. These exist for both coaxial line (see Fig. 11.1) and waveguide. The method depends upon the measurement of the positions of the standing-wave mimima and the voltage standing-wave ratio (VSWR) produced by the unknown.

The equipment used for slotted line measurements is illustrated in Fig. 11.9. The signal picked up by the probe on the slotted line is passed from the detector diode to a special instrument known as a VSWR meter. The oscillator is square-wave modulated (normally at 1 kHz.) and the VSWR meter incorporates a tuned amplifier which rejects all signals outside a narrow band centred on the modulation frequency. The purpose of this is to give the best possible signal-to-noise ratio. The VSWR meter also contains both step and infinitely variable attenuators which are used to set the level displayed on the meter and to change from range to range. The meter is calibrated directly in VSWR and usually has scales displaying decibels as well. The way in which measurements are made is described below.

If the reflection coefficient of the unknown is ϱ then the reflected wave amplitude is

$$V_r = \varrho V_i. \tag{11.6}$$

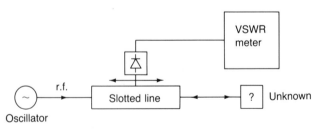

Fig. 11.9 Block diagram of equipment for making slotted line measurements of impedance.

The maximum of the standing wave has amplitude

$$V_{\max} = V_i(1 + |\varrho|) \tag{11.7}$$

and the minimum

$$V_{\min} = V_i(1 - |\varrho|) \tag{11.8}$$

so that the voltage standing-wave ratio is

$$S = \frac{1 + |\varrho|}{1 - |\varrho|}. \tag{11.9}$$

The VSWR is measured by moving the probe to a signal maximum and setting the meter to full-scale deflection (marked '1' on the VSWR scale) using the attenuators. The probe is then moved to a signal minimum and the VSWR read directly from the meter. Greater accuracy can be obtained by taking the average of several measurements.

By rearranging (11.9) we find that the magnitude of the reflection coefficient is given by

$$|\varrho| = \frac{S - 1}{S + 1}. \tag{11.10}$$

The equipment can therefore be used to make measurements of reflection coefficient as an alternative to a reflectometer. It has the disadvantage that the measurements must be made at spot frequencies.

The positions of the minima can be used to determine the magnitude and phase angle of the impedance at a reference plane. At a standing-wave minimum the current is given by

$$I_{\min} = \frac{V_i}{Z_0} (1 + |\varrho|) \tag{11.11}$$

so that the apparent impedance of the unknown at that plane is

$$Z'_L = V_{\min}/I_{\min}$$
$$= Z_0\left(\frac{1 - |\varrho|}{1 + |\varrho|}\right)$$
$$= Z_0/S. \tag{11.12}$$

Thus, at a standing wave minimum, the apparent impedance is real.

If the unknown is replaced by a short circuit as shown in Fig. 11.10 a new set of minima can be detected which are spaced at half wavelength intervals from the short circuit. Suppose that there is a minimum at P in the presence of the unknown and one at P' with the short circuit. Then, because the apparent impedance of the unknown is known at P, it can be calculated at P' using the usual formula for transformation of impedance on a transmission line (Carter, 1986)

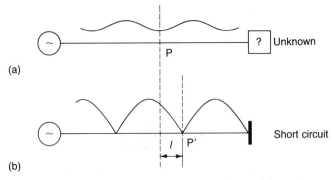

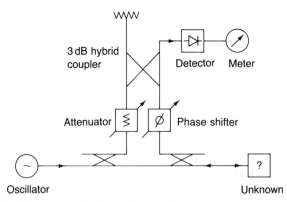

Fig. 11.10 Measurement of impedance using a slotted line: (a) standing wave when the line is terminated by the unknown, and (b) standing-wave pattern with a short-circuit termination.

$$\frac{Z_{p'}}{Z_0} = \frac{Z_p + jZ_0 \tan kl}{jZ_p \tan kl + Z_0},\tag{11.13}$$

where l is positive if P′ lies closer to the generator than P. The transformation of impedance can be carried out graphically using a Smith chart (see Appendix A).

Although, in theory, the maxima of the standing wave could be used for these measurements the minima are always used because their positions can be found more accurately.

Another way of measuring the phase angle of a reflected signal is to use a phase bridge. The arrangement of the bridge is shown in Fig. 11.11. The reflected signal is sampled by a directional coupler and passed through a calibrated variable phase shifter. The incident signal is sampled likewise

Fig. 11.11 Block diagram of a phase bridge for measuring the phase angle of the reflection coefficient of the unknown.

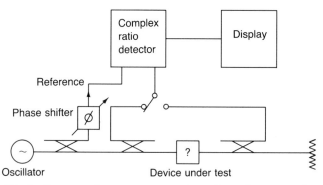

Fig. 11.12 Block diagram of a vector network analyser for measuring complex transmission and reflection coefficients.

and passed through a variable attenuator. These two signals are then combined using a 3 dB hybrid coupler so that the phasor sum is detected. The attenuator is adjusted so that the amplitudes of the signals are the same. Then the phase shifter is adjusted to produce a signal null at the detector. If the unknown is replaced by a short circuit the measurement can be repeated to give a phase reference.

The phase bridge just described is an r.f. substitution method. It has largely been supplanted by an instrument in which the reference and reflected signals are converted to an intermediate frequency at which the phase comparison is made. This system, known as a vector network analyser, is shown in Fig. 11.12. The signals reflected from or transmitted by the unknown are compared with a reference signal by a complex ratio detector. The phase of the reference signal is adjusted by a phase shifter. A modern vector analyser usually contains an accurate signal source and a computer which is able to carry out error correction and calibration as well as displaying the results of the measurements in a variety of forms. For a more detailed discussion of the vector network analyser see Bryant (1988).

11.7 TIME-DOMAIN REFLECTOMETRY

In a complex microwave system it is sometimes easier to measure the mismatch at a port than to say exactly what part of the system is responsible for it. Time-domain reflectometry (TDR) provides a complementary technique to the frequency-domain reflectometry described in Section 11.5. The equipment used is illustrated in Fig. 11.13.

The step generator applies a step function to the system under test and the voltage on the input line is detected with a high-impedance probe. If the system is perfectly matched then the probe voltage is just half the generator voltage. The presence of mismatches within the system causes pulses of

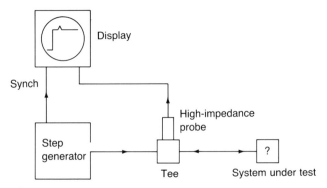

Fig. 11.13 Block diagram of equipment for making measurements by time-domain reflectometry.

returned power which add to, or subtract from, the input pulse depending upon the sign of the reflection (Carter, 1988, p. 113 ff.). The time delay of the returned pulse gives a measure of the position of the mismatch within the system. This technique is used at lower frequencies to find faults in telephone cables. If the rise time of the pulse is very fast (of the order of a picosecond) then it is possible to resolve discontinuities which are a few millimetres apart.

The method just described can only be used with systems which use two conductor lines so that they can carry signals down to d.c. For systems of limited bandwidth such as waveguides it is necessary to use pulsed microwave signals instead (Gardiol, 1984).

The time-domain and frequency-domain responses of a circuit are related to one another (Dunlop and Smith, 1984, Ch. 1). Suppose that the reflection coefficient measured at the circuit input as a function of frequency is $\varrho(\omega)$ and that the input signal is $V_i(t)$. The input signal can be expressed as the superposition of sinusoidal waves whose amplitude varies with frequency as

$$G(\omega) = \frac{1}{2\pi} \int_{-\infty}^{\infty} V_i(t)e^{-j\omega t}\, dt. \tag{11.14}$$

The amplitude of the reflected signal at any frequency is obtained by multiplying $G(\omega)$ by the reflection coefficient. Finally the system response in the time domain is obtained by Fourier synthesis as

$$f(t) = \int_{-\infty}^{\infty} G(\omega)\varrho(\omega)e^{j\omega t}\, d\omega. \tag{11.15}$$

The relationship between the time- and frequency-domain descriptions of the reflection can be used to compute the one from the other. Some vector network analysers can compute the TDR response from a swept frequency measurement using the fast Fourier transform (FFT) method (Dunlop and Smith, 1984, p. 21).

11.8 SPECTRUM ANALYSER MEASUREMENTS

When measurements are to be made on non-linear or active microwave systems the waveforms may not be sinusoidal. This may be because they are amplitude or frequency modulated, or contain harmonics or spurious frequencies caused by parasitic oscillations. In all these cases it is useful to be able to take the actual waveform in the time domain and analyse it into its frequency components. An instrument which performs this function is called a spectrum analyser.

In principle spectrum analysis could be carried out using a tunable narrow-band filter. The output from the filter would then be proportional to the harmonic amplitude at each frequency. In practice it is difficult to make such a filter except for small frequency ranges. An alternative is to mix the signal to be analysed with that of a swept-frequency local oscillator as shown in Fig. 11.14. The result is the production of sum and difference frequencies of which the latter is selected by a narrow-band i.f. amplifier and passed through an envelope detector to the display. There can be problems with this arrangement if the swept frequency has appreciable harmonic content. The output from the mixer would then be

$$f_i = f \pm n f_0 \tag{11.16}$$

so that the same f_i could be generated by more than one input frequency depending upon the value of n. A possible solution to this problem is to pass the input signal through a swept filter before it enters the mixer. This filter can be quite broad band because it only needs to eliminate those frequencies which are far enough from the centre frequency to produce spurious output. Thus if the analyser is designed to work with the nth harmonic of the swept oscillator the nearest frequencies which can give spurious output are

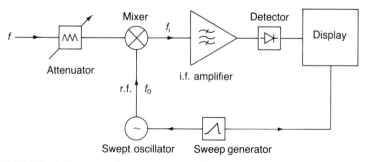

Fig. 11.14 Block diagram of a spectrum analyser.

$$f = f_i + (n + 1)f_0 \tag{11.17}$$

and
$$f = f_i + (n - 1)f_0 \tag{11.18}$$

so that the pre-filter bandwidth must be less than $2f_0$.

11.9 ELECTROMAGNETIC COMPATIBILITY MEASUREMENTS

Electromagnetic compatibility measurements fall into two main classes: the susceptibility of equipment to external electromagnetic fields and the emission of radiation from equipment.

Measurements of immunity to external fields are made by placing the equipment under test (EUT) in a known radiation environment. Field strengths as high as $200\,\mathrm{V\,m^{-1}}$ are required by military test specifications. At low frequencies this can be achieved by placing the EUT between the conductors of a specially constructed section of TEM transmission line. Figure 11.15 shows the arrangement of one such TEM cell. A section of $50\,\Omega$ parallel-plate line is connected by tapers to coaxial lines. This arrangement is satisfactory provided that the dimensions of the EUT do not exceed 30% of those of the cell and that the width of the line is less than half a wavelength. The method is used at frequencies up to $500\,\mathrm{MHz}$.

At microwave frequencies the test field is provided by antennas of known gain radiating within anechoic chambers. There is, however, the difficulty that the field is perturbed by the presence of the EUT since the latter

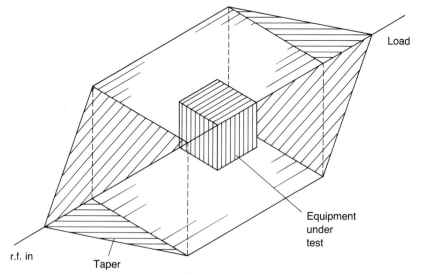

Fig. 11.15 Arrangement of a TEM cell for measuring the electromagnetic susceptibilities of electronic equipment.

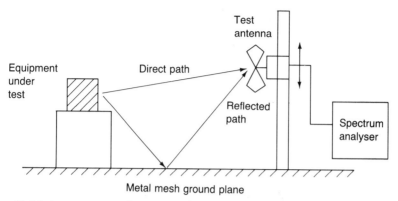

Fig. 11.16 Arrangement of an open field test site for measuring the radiation emitted from electronic equipment.

commonly has a metal case. Thus it is necessary to specify whether the test field is to be set up in the presence or absence of the EUT.

Radiated emission measurements are commonly made on an open field test site as shown in Fig. 11.16. The radiation from the EUT is received by a standard antenna and fed to a spectrum analyser. Because reflection of the radiation from the ground is unavoidable a metal mesh ground plane is used to ensure that the reflection is as close to that predicted by theory as possible. When the test antenna is moved vertically on its mast maxima and minima of the signal are observed corresponding to constructive and destructive interference between the direct and reflected signals.

Open field test sites suffer from the disadvantage that the test antenna also picks up the signals from nearby radio transmitters. To avoid this problem an anechoic chamber can be used, but it is difficult to make one which is satisfactory below 100 MHz. At lower frequencies screened rooms have been used to make measurements of radiated emission. In order to do this satisfactorily it is necessary to characterize the room at all frequencies in the range of interest because of the effects of reflections from the walls. An added problem with low-frequency measurements is that the receiving antenna is inevitably in the induction field rather than the radiation field of the EUT.

In many cases emission or susceptibility of the EUT has more to do with conduction along cables rather than direct radiation. Measurements are then made by using current transformers to couple signals off or on to the cables as shown in Fig. 11.17. The ferrite rings provide matched terminations for the signals to be measured.

Electromagnetic compatibility measurements are becoming much more important because of the introduction in 1992 of an EEC directive that all electrical equipment marketed and put to use within the EEC must satisfy

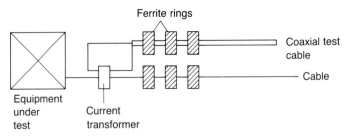

Fig. 11.17 Schematic diagram showing the use of a current transformer and ferrite damping rings to measure the conducted susceptibility and emission of electronic equipment.

regulations for both emission and immunity. Further information is given by Keiser (1983) and Jackson (1989).

11.10 MEASUREMENT OF RESONATORS

A microwave resonator is characterized by its resonant frequency, Q factor and shunt impedance (R/Q). These parameters can be measured using the equipment illustrated in Fig. 11.18. The signal from a microwave sweep oscillator is coupled into the resonator by a probe. The coupling must be weak to avoid loading of the resonator by the measuring system. A similar probe couples the output to a scalar analyser. The frequency of the oscillator is measured by coupling some of the signal into a microwave counter. Because the counter requires an unmodulated input and the scalar analyser requires a modulated input it is necessary to modulate the output signal from the resonator before it is detected.

If the oscillator is swept over a wide band of frequencies a series of peaks appears on the screen of the scalar analyser corresponding to the different modes of the resonator. The oscillator signal must not have appreciable

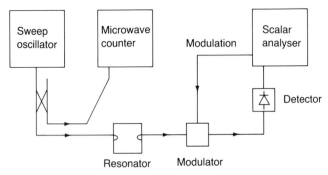

Fig. 11.18 Block diagram of equipment for measuring the properties of microwave resonators.

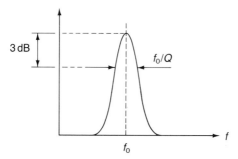

Fig. 11.19 Measuring the Q of a resonator from the width of its frequency-response curve.

harmonic content or spurious responses will appear. The sweep can then be narrowed down to isolate a single resonance (see Fig. 11.19) and the frequency swept manually to locate the peak of the response. This frequency is read directly from the counter to an accuracy of around $0.1/Q$. The vertical scale of the scalar analyser display is calibrated in decibels so it is easy to locate the 3 dB points, measure their frequencies, and calculate the Q factor of the resonator from (7.10).

To measure the shunt impedance of a cavity resonator a perturbation technique is used. This can either involve a perturbation of the cavity boundary by a metal probe or the insertion of a dielectric rod (Waldron, 1967). The latter technique will be described here with reference to the TM_{01} mode of a cylindrical pillbox cavity.

Figure 11.20(a) shows such a cavity with a thin dielectric rod inserted along its axis. For this mode of oscillation the electric field is axial and maximum on the axis. Provided that the perturbing effect of the rod is small we can assume that the fields in the cavity outside the rod are unaffected by its presence. The equivalent circuit of the perturbed cavity is then as shown in Fig. 11.20(b) where L_0 and C_0 are the inductance and capacitance of the unperturbed cavity and C_1 represents the effect of the dielectric rod. C_1 is calculated by treating the rod as a parallel-plate capacitor

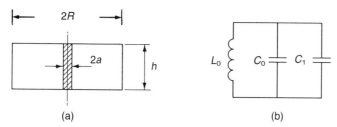

Fig. 11.20 Perturbation of a cylindrical microwave resonant cavity by a dielectric rod: (a) general arrangement, and (b) equivalent circuit.

and subtracting the capacitance of an air-spaced capacitor having the same shape. The result is

$$C_1 = \frac{\varepsilon_0 \pi a^2}{h} (\varepsilon_r - 1), \tag{11.19}$$

where ε_r is the relative permittivity of the rod. If the frequency perturbation is $\Delta\omega$ then, since the frequency is proportional to $1/\sqrt{C}$,

$$\frac{\omega_0 + \Delta\omega}{\omega_0} = \sqrt{\left(\frac{C_0}{C_0 + C_1}\right)}. \tag{11.20}$$

If $C_1 \ll C_0$ the square root can be expanded by the binomial theorem to give

$$\frac{\Delta\omega}{\omega_0} \simeq -\frac{C_1}{2C_0}. \tag{11.21}$$

Substituting for C_1 from (11.19) gives

$$\frac{\Delta\omega}{\omega_0} \simeq -\frac{\varepsilon_0 \pi a^2 (\varepsilon_r - 1)}{2hC_0}. \tag{11.22}$$

Now from (7.4)

$$\frac{1}{C_0} = \omega_0 \left(\frac{R}{Q}\right) \tag{11.23}$$

so that (11.22) becomes

$$\frac{\Delta\omega}{\omega_0^2} \simeq -\frac{\varepsilon_0 \pi a^2 (\varepsilon_r - 1)}{2h} \left(\frac{R}{Q}\right). \tag{11.24}$$

Thus, if the dimensions of the rod and its relative permittivity are known the R/Q of the cavity can be calculated from measurements of ω_0 and $\Delta\omega$. Note that the negative sign in (11.24) means that the frequency is perturbed downwards by the rod.

A more exact analysis takes account of the changes in the cavity fields as a result of the presence of the rod. For a ratio $a/R = 0.06$ the error in R/Q resulting from the use of the approximate formula (11.24) is about 3%.

This technique is used to measure the R/Q of cavities for klystrons and can be adapted to measure the coupling impedances of the slow-wave structures used in travelling-wave tubes and linear accelerators (Ch. 10) (Connolly, 1976).

11.11 MEASUREMENT OF DIELECTRIC PROPERTIES

The dielectric properties of a material can be measured by essentially the same method as that described for measuring R/Q in the previous section.

If the cavity resonator has a simple shape such as a cylindrial pillbox then its R/Q can be calculated from theory (see Section 7.3). Equation (11.24) can then be used to calculate the relative permittivity of the perturbing rod from the frequency shift. The change in the Q of the cavity produced by the rod can be used to calculate the loss tangent of the material.

This method suffers from a number of disadvantages. It requires the sample of material to be of a particular size and shape, it suffers from errors caused by the approximations made in deriving equation (11.24) and it is not good for measuring the loss tangents of low-loss materials because the change in Q is too small to measure accurately. The first two problems can be overcome by using a method in which the sample protrudes into a cavity through a slot and making measurements for different depths of insertion. When the frequency shift is plotted against the insertion depth the slope of the line gives the permittivity free from errors produced by the finite size of the sample and the slot in the cavity wall through which it protrudes.

Another technique for measuring the dielectric properties of materials depends on measuring the properties of a section of coaxial line or waveguide which is filled with the material. The transmission loss and return

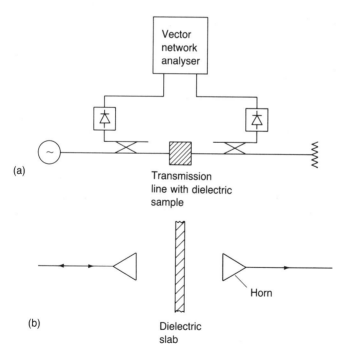

Transmission
line with dielectric
sample

(a)

(b)

Horn

Dielectric
slab

Fig. 11.21 Measurement of the dielectric properties of materials: (a) using a section of transmission line filled with the dielectric, and (b) using a slab of dielectric.

loss can be measured with a vector network analyser (see Fig. 11.21(a)) and the dielectric properties of the sample deduced.

When the sample of the material is in the form of a large slab its properties can be measured by placing the slab between two microwave horns as shown in Fig. 11.21(b). The slab has the effect of changing the path length between the horns and also of reflecting some of the incident power because of the mismatch of wave impedance at its surface. The transmitted and relected signals are then analysed and the dielectric properties calculated as in the previous paragraph.

A useful review of methods of measuring dielectric properties and a table giving figures for a wide range of materials are given by Metaxas and Meredith (1983).

11.12 CONCLUSION

This chapter has set out to show how the devices and theory discussed earlier in the book are employed in a variety of microwave measuring systems. Most of the techniques and instruments likely to be found in an industrial or university microwave laboratory have been described. For the most part they are used routinely as part of the process of product development or manufacture. Very often the instruments are connected to each other and to a computer by a data bus which enables the measurements to be automated and results output which have been calculated from the raw data. Other, more specialized, techniques exist which are used in standards laboratories and for measurements in physics laboratories.

When a measurement is made it is important to know the sources of possible error. The measuring technique must minimize these as far as possible and provide an estimate of the residual error. In this book it has only been possible to give a brief discussion of this subject. It is particularly important to remember that sophisticated modern instruments which appear to give very accurate results are not free from errors. The manufacturers' manuals usually include discussions of measurement errors and their correction and these should be studied carefully. Fuller discussions of microwave measuring techniques including the signal-flow graph method for estimating errors will be found in the books by Bryant (1988) and Laverghetta (1976).

EXERCISES

11.1 When a slotted section of air-spaced coaxial line is terminated by a short circuit the standing-wave minima are 18.6 mm, 16.3 mm, 12.7 mm and 9.6 mm apart at four different frequencies. What are those frequencies?

11.2 If the figures given in Question 11.1 were obtained with a WG16 waveguide slotted line what would the frequencies be?

11.3 Examine the effect of increasing the source reflection coefficient in the example on p. 269 to 0.1.

11.4 In a series of slotted-line measurements the separation of the minima is 36.5 mm. When the unknown loads are replaced by a short circuit the minima move towards it by 18.2 mm, 23.7 mm and 31.8 mm. The corresponding VSWR figures are 1.07, 1.20, 1.43. Find the normalized impedances of the unknown loads.

11.5 A cylindrically symmetrical cavity resonator is 11 mm high and resonates at 3.56 GHz with the electric field directed vertically. When a 3 mm glass rod ($\varepsilon_r = 4.1$) is inserted along the axis the resonant frequency drops by 45 MHz. Calculate the R/Q of the cavity.

11.6 The resonator described in Question 11.5 is perturbed by a different dielectric rod 2.54 mm in diameter. If the frequency drop is 94 MHz what is the relative permittivity of the rod?

<table>
<tr><td>12</td><td># Systems using electromagnetic waves</td></tr>
</table>

12.1 INTRODUCTION

The purpose of this chapter is to provide a brief survey of the range of systems which depend upon electromagnetic waves for their operation. They fall into two main categories: those which are concerned with the transmission of information and those whose purpose is the transmission of power. The first category includes radio, television, satellite communications and radar. The second embraces industrial and domestic microwave and r.f. heating, medical hyperthermia and schemes for using microwaves to beam power down from satellite power stations. The total range of systems is so diverse that it is only possible in this book to give some idea of the main applications of electromagnetic waves.

12.2 RADIO WAVE PROPAGATION

Electromagnetic waves travel in a straight line in a uniform medium so it might be expected that they could only be used for line-of-sight communications. In fact, as is well known, worldwide communication is possible at some frequencies. The various mechanisms responsible for the propagation of radio waves are illustrated in Fig. 12.1. The layers of the atmosphere which are involved are the troposphere and the ionosphere. The former is the layer lying closest to the surface of the Earth with a typical thickness of 10 km. Most of the weather variations in the atmosphere take place in this layer. The ionosphere is made up of a series of layers between 50 km and 400 km from the Earth's surface which are characterized by the presence of free electrons generated by ionizing radiation.

The assumption that the waves travel in straight lines is only correct if they are moving through a uniform medium. Because the density of the atmosphere decreases with height so, therefore, does its refractive index. Equation (1.91) shows that the effect of this decrease is to cause rays to be

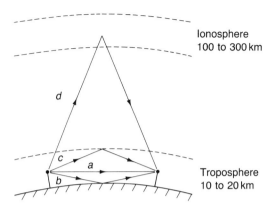

Fig. 12.1 Paths between transmitter and receiver for radio waves: (a) direct wave, (b) reflected wave (known together as the ground wave), (c) tropospheric wave and (d) sky wave.

bent away from the normal as they travel upwards and towards the normal as they travel downwards. The refractive index of the atmosphere, even at sea level, is close to unity (typically 1.0003) so that the effect is very slight. Nevertheless, over long distances the curvature of the path a in Fig. 12.1 is appreciable. Figure 12.2 shows the effect of this on communication between two antennas which are elevated above the Earth's surface. If the path between the two antennas is a straight line then the point B is the furthest point from the transmitter at which satisfactory reception is possible. The curvature of the ray actually makes it possible for the signal to be received at B'. Calculations involving curved paths are inconvenient so it is usual to allow for the curvature of the path AB' by using an effective radius for the Earth which is about 4/3 of the actual radius. It must be remembered that the properties of the atmosphere vary from time to time so that communication over the horizon which relies on tropospheric refraction is not always certain.

Propagation through the atmosphere is affected by several different processes of absorption. At frequencies above 10 GHz absorption by the molecules of the atmosphere is important. Water molecules have a resonance at 22 GHz which produces an attenuation of about 0.16 dB km^{-1}. The oxygen molecule has a group of absorption lines at 60 GHz which

Fig. 12.2 Showing how the range of radio communication is increased by tropospheric refraction.

produce an attenuation of 15 dB km^{-1}. The actual attenuation depends on the atmospheric pressure and humidity. For signals travelling vertically through the atmosphere in a satellite communications system the one-way attenuation from these mechanisms is less than 1 dB for frequencies up to 50 GHz. Attenuation by rainfall increases with frequency and with the amount of rain. At 10 GHz the attenuation is around 1 dB km^{-1} when the rainfall is 25 mm h^{-1}.

Besides the direct wave from transmitter to receiver there will normally also be a reflected wave (b in Fig. 12.1). To examine this effect consider the situation shown in Fig. 12.3. Let us assume for simplicity that the transmitting antenna is a small horizontal dipole and that the curvature of the Earth can be neglected. The electric field at the receiving antenna, B, can be expressed as

$$E_B = E_A + E_{A'} \qquad (12.1)$$

where E_A and $E_{A'}$ are the electric field intensities at B for waves taking the paths AB and ACB respectively. The reflected wave can be regarded as being generated by an image antenna A' whose amplitude and phase are adjusted to allow for the finite conductivity of the Earth's surface. Then, by a straightforward extension of the theory of Section 5.8 the field at B is

$$E_B = E_A \exp j\psi/2 + \varrho E_A \exp -j\psi/2, \qquad (12.2)$$

where ϱ is the amplitude of the reflection coefficient of the Earth,

$$\psi = 2k_0 h_1 \cos \phi + \alpha \qquad (12.3)$$

and α is the phase difference between antenna A and its image A'.

If the Earth is regarded as a perfect conductor then $\varrho = 1$ and $\alpha = 180°$. The field at B is then

$$\begin{aligned} E_B &= 2E_A \cos (k_0 h_1 \cos \phi + \pi/2) \\ &= 2E_A \sin (k_0 h_1 \cos \phi) \\ &= E_A F, \end{aligned} \qquad (12.4)$$

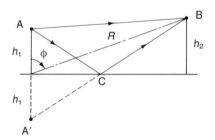

Fig. 12.3 Geometry of ground-wave reception.

where

$$F = 2 \sin \left(\frac{k_0 h_1 h_2}{R} \right). \tag{12.5}$$

The factor F is known as the path gain factor. It can vary from zero to 2 depending upon the relative positions of the antennas. Rather more complicated expressions for this factor emerge if the finite conductivity and curvature of the Earth are taken into account. The principle remains the same, namely that the field at the receiving antenna varies depending upon whether the interference between the direct and reflected waves is constructive or destructive. Radio sets often include an automatic gain control (AGC) loop so that this effect is not apparent to the user. If, however, the radio is operating at the limit of its sensitivity then the AGC loop can no longer compensate for the variations in the signal received. Thus the reception of a weak station by a car radio varies noticeably as the car moves along the road.

The reflection of waves by the Earth's surface is further complicated by scattering and diffraction of the waves by hills, trees and buildings. These effects become important at frequencies above 10 MHz (30 m wavelength) when the typical size of the obstacles is comparable with the wavelength of the waves. At lower frequencies the effects can be included in the effective reflection coefficient of the ground.

The upper limit of the troposphere is marked by a temperature inversion (warm air above cold air) with an abrupt change in the refractive index. The refractive index is also affected by the humidity of the air. Any abrupt change in refractive index will reflect radio waves so that they follow a path such as c in Fig. 12.1. Tropospheric reflection can increase the range of reception to a hundred miles or more beyond the geometrical horizon. Under certain special conditions it is possible for the curvature of the ray to be equal to that of the Earth's surface. This effect, known as ducting, can extend the range of communication to several thousand miles and can intefere with microwave communication links.

The effects described in the previous paragraph are strongly dependent upon the atmospheric conditions which exist in the troposphere at a particular time. They are, therefore, an uncertain means of long-range communication. The reflection of waves by the ionosphere which is more constant (though still variable) is normally used for long-range radio. As ionizing particles and radiation enter the atmosphere they pass freely until the air density is high enough for appreciable ionization to take place. There is then a region of increasing ionization until all the particles and photons have given up their energy. The resulting distribution of charge density with height is shown in Fig. 12.4. The ionosphere has a number of layers with differing properties as shown in Fig. 12.4. Of these the most important for radio communication is the F_2 layer whose ionization varies

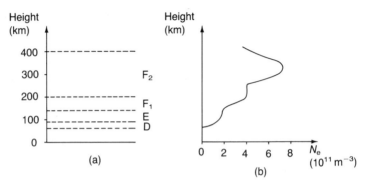

Fig. 12.4 The ionosphere: (a) approximate heights and, (b) approximate electron densities of the layers.

in a complex way linked to the sunspot number. The F_1 layer is the result of daytime solar radiation so it disappears, at night being merged with the F_2 layer. The E region is ionized by solar ultraviolet and X-rays and, to some extent by cosmic rays. The D region has much lower ionization and its main effect is to attenuate waves passing through it.

The propagation of waves through an ionized gas was studied in section 1.6 where it was shown that the permittivity is given by (1.62)

$$\varepsilon = \varepsilon_0\left(1 - \frac{\omega_p^2}{\omega^2}\right). \tag{12.6}$$

Thus the refractive index of the F_2 layer ($n = \sqrt{\varepsilon}$) is less than that of free space and waves are therefore reflected by it. Strictly, the effects of collisions in the electron plasma should be included in the theory. The result is that the permittivity is given by

$$\varepsilon = \varepsilon_0\left[1 - \frac{\omega_p^2}{(\omega^2 + v^2)}\right], \tag{12.7}$$

where v is the mean collision frequency. The energy dissipation by the collisions results in an effective conductivity

$$\sigma = \frac{\omega_p^2 v \varepsilon_0}{\omega^2 + v^2}. \tag{12.8}$$

The conductivity depends upon both the electron density and the collision frequency. At high altitudes v is very low whilst ω_p decreases rapidly below the bottom of the E layer. It turns out that the conductive effects are limited to a thin layer at the bottom of the E region. This layer therefore causes high attenuation of signals passing through it.

The reflection of waves by the ionosphere varies with frequency. At frequencies below 150 kHz the change in the electron density within one wavelength is so great that it can be represented as a step in the refractive index so that the wave is reflected abruptly. The propagation of waves at low frequencies is analysed in terms of waveguide modes propagating between the Earth's surface and the ionosphere. These frequencies are normally used for worldwide communication and for navigation systems such as Loran C.

At higher frequencies it is necessary to regard the wave as passing through a medium whose refractive index varies with position. The result, as with tropospheric propagation, is that the ray becomes curved and may be turned back towards the Earth's surface. Frequencies above 30 MHz are not very well reflected by the ionosphere and cannot be used for long-distance communications. Short-wave radio (3 to 30 MHz) provides useful, medium-power, long-range communications in which the whole of the received signal is reflected from the ionosphere. Because of the variable nature of the ionosphere it is subject to fading, multipath interference and high noise interference.

In the 150 to 1500 kHz band the ground wave is stable and subject only to moderate attenuation and the ionospheric reflection at night is reliable. This band is widely used for marine communications and navigation and medium-wave broadcasting. The signal received is made up of the sum of the ground wave and the sky wave. It is therefore subject to multipath interference and fading. Wave propagation in the ionosphere is subject to Faraday rotation because of the effects of the Earth's magnetic field (see Section 1.8) and careful attention must be paid to the polarization of the antennas if good reception is to be obtained.

It will be evident that this section is a very brief summary of a very complex subject. For fuller information consult Jordan and Balmain (1968) and Kirby (1982).

12.3 RADIO COMMUNICATIONS

The earliest use of electromagnetic waves was in radio communications. From its beginnings as wireless telegraphy this field has expanded to include public broadcasting at frequencies from 'long wave' (150 kHz) to VHF (100 MHz) and mobile and cellular radio. We shall not discuss here the variety of modulation schemes used to maximize the signal-to-noise ratio or to make best use of the available bandwidth. The reader should consult books on telecommunications such as Dunlop and Smith (1984) for further information on this subject. We note in passing that the frequency bandwidth required for a given flow of information is independent of the carrier frequency and that many more communication channels can be fitted into the range 90 to 100 MHz than into the range 150 to 300 kHz.

Figure 12.5 shows a simple short-range radio communication system. The two antennas are assumed to be in each other's far field but close

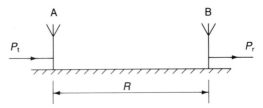

Fig. 12.5 A simple short-range radio communication system.

enough together so that the curvature of the Earth's surface can be neglected. If, for the moment, we ignore the possibility that waves are reflected from the surface of the earth then the relationship between the transmitted power and the received power is given by (5.48)

$$P_r = \frac{A_{e2}g_1}{4\pi R^2} P_t,$$
(12.9)

where A_e is the effective area of an antenna and g is its gain. We also saw in Chapter 5 that these two quantities are related to each other by (5.54)

$$A_e = \frac{\lambda^2}{4\pi} g$$
(12.10)

so that (12.9) can be rewritten in the form

$$\frac{P_r}{P_t} = \frac{A_{e1}A_{e2}}{R^2\lambda^2}.$$
(12.11)

This equation is known as the Friis transmission formula. It can also be written as

$$\frac{P_r}{P_t} = \frac{g_1 g_2 \lambda^2}{(4\pi R)^2}.$$
(12.12)

Equation (12.12) reveals some basic facts about radio communications. If we assume that a certain P_r is the minimum which will ensure that the signal received is above the noise level then doubling the range requires the transmitted power to be increased by a factor of 4 if the same antennas are used. Alternatively the gains of the antennas could be increased to compensate for the increased range by increasing either their size or their directivity. It appears from (12.12) that the received power should increase as the wavelength increases. In general this does not happen because the directivities of antennas tend to be lower at longer wavelengths.

It is often convenient to express (12.12) in decibels with the result

$$P_r = P_t + G_1 + G_2 - L_s,$$
(12.13)

where the transmitted and received powers (P_t and P_r) and the antenna gains (G_1 and G_2) are expressed in decibels and the free-space path loss L_s is given by

$$L_s = 20 \log_{10} \left(\frac{4\pi R}{\lambda} \right). \tag{12.14}$$

In any real communication system there must be other losses in the feeder cables and in propagation through the air. These losses, which may be variable to take account of rainfall, can be included as a propagation loss L_p to give

$$P_r = P_t + G_1 + G_2 - L_s - L_p. \tag{12.15}$$

Careful control over the use of the available frequency spectrum is exercised by national regulatory bodies with the aim of minimizing interference between stations. Frequencies are allocated for international, national or local use with prescribed maximum power levels. Even with these precautions it is common experience that there can be interference between stations particularly in the medium wave band at night when the ionospheric reflections are at their best. Transmitted power levels of up to 50 kW are normal for national broadcasting with greater powers being used for international short-wave stations.

Radio communication at frequencies above 30 MHz is virtually limited to line-of-sight paths with the requirement becoming more stringent as the frequency increases and diffraction effects are less helpful in providing a signal in the geometrical shadow of an obstacle. Because waves at these frequencies are not affected by ionospheric reflections it is possible to use horizontal and vertical polarization to provide freedom of interference between neighbouring stations which are transmitting at the same frequency.

12.4 TELEVISION BROADCASTING

Colour television broadcasting requires a bandwidth of 8 MHz for each channel. Channels in the UHF region (470 to 890 MHz) are allocated for TV transmission because there is adequate bandwidth available for a large number of stations whilst the receiver technology is less expensive than at higher, microwave, frequencies. At UHF there must be a line of sight between the transmitting and receiving antennas. Main transmitters employ very tall aerial masts and transmit powers of up to 50 kW. Additional local transmitters are needed to provide coverage in areas shadowed by hills. The polarization of the wave is used to avoid inter-ference between adjacent transmitters.

At UHF it is possible to make receiving antennas with gains greater than 10 dB which are compact enough for domestic use. The transmitting antennas make use of dipole arrays which are usually arranged to give uniform coverage in all directions. The phases of the feeds to dipoles at different heights on the mast are adjusted to tilt the radiation pattern down towards the horizon. In some cases the antenna may also be arranged to give an asymmetrical radiation pattern in the horizontal plane.

12.5 MICROWAVE COMMUNICATIONS

An important use of microwave radio is in point-to-point communication links of the kind typified by the Telecom Tower in London and similar towers elsewhere. There are a number of frequency bands allocated for this purpose of which the most used are those in the region of 4 and 6 GHz, each of which has 500 MHz bandwidth. The route chosen must be strictly line-of-sight with stages typically 40 km in length. The antennas used have high gain (40 to 50 dB) and consequently low side-lobe powers. The result is that such a communications system has a good level of security. The number of channels which can be handled by the system is doubled by making use of both horizontal and vertical polarization. This kind of system provides a good example of the use of equation (12.15).

Example

Find the power received at the end of a 40 km, 6 GHz, microwave link if the antennas have 40 dB gain, the transmitted power is 20 W and the propagation losses are not more than 6 dB.

Solution

The wavelength at 6 GHz is 50 mm so the free space loss is

$$L_s = 20 \log_{10} \left(\frac{4\pi \times 40\,000}{0.05} \right) = 140\,\text{dB} \tag{12.16}$$

and the transmitted power is

$$P_t = 10 \log_{10} \left(\frac{20}{0.001} \right) = 43\,\text{dBm} \tag{12.17}$$

so that the minimum power received is

$$\begin{aligned} P_r &= 43 + 40 + 40 - 140 - 6 \\ &= -23\,\text{dBm} \end{aligned} \tag{12.18}$$

that is 5 μW. This signal level is comfortably within the range of sensitivity of microwave receivers.

If the strength of the signal received is plotted against the distance between the transmitter and receiver the result is as shown in Fig. 12.6. As would be expected the signal level falls off steadily with periodic variations caused by interference between the direct and reflected waves. Once the transmitter is over the horizon as perceived by the receiver the signal level falls off very rapidly. The signal does not fall abruptly to zero because of the effects of diffraction. This can be understood by considering Fig. 12.7 which shows a receiver whose line of sight to the transmitter is blocked by a hill. The radiation from the transmitter illuminates the plane A–A containing the hill. The radiation at any point beyond this plane can be calculated by using Huygens' method of secondary wavelets (see Section 5.9). The result is that the intensity of the radiation at points on the plane B–B varies as shown in Fig. 12.8 (Longhurst, 1973). Within the geometrical shadow there is still some detectable signal because the effects of the individual wavelets do not quite cancel out.

The existence of the diffraction field can be used for over-the-horizon communications. It is sometimes possible to make use of diffraction by a mountain to extend the range of a microwave link. The penalty to be paid

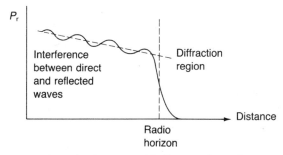

Fig. 12.6 Variation of received power with distance from the transmitter.

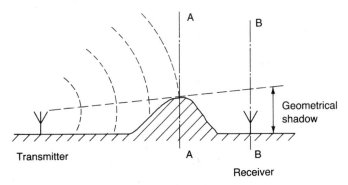

Fig. 12.7 Geometry of reception beyond an obstacle.

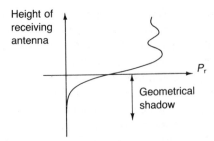

Fig. 12.8 Variation of received power caused by diffraction by an obstacle.

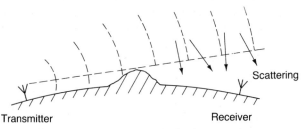

Fig. 12.9 Increased signal level at an over-the horizon receiver as a result of tropospheric scattering.

is the need for much greater transmitter power than for a line-of-sight link. As the receiver moves further from the obstacle and, therefore, further into the geometrical shadow it is found that the signal level does not fall off as fast as predicted by theory. This is explained as the result of scattering of the radiation by turbulence and inhomogeneity in the troposphere as shown in Fig. 12.9. The effect is put to use in troposcatter communication systems which are useful when it is not possible to install repeater stations. An example is the system used to communicate with oil rigs in the North Sea. A transmitter power of the order of 1 kW is needed to provide a strong enough signal at the receiver.

Satellite communication links are a special kind of microwave link. The satellite is normally in geostationary orbit at a radius of 35 800 km. This arrangement has the advantage that the satellite is always in view of the Earth station and that no Doppler shifting of the frequency is caused by motion of the satellite relative to the Earth. On the other hand it places greater demands upon the microwave system than a satellite in a lower orbit (because of the very large range) and introduces a propagation delay of about 0.5 s into a two-way telephone conversation.

Many current systems use frequencies around 6 GHz for the uplinks and around 4 GHz for the downlinks. The free-space path loss is 200 dB for a satellite in geostationary orbit at these frequencies so it is necessary to use high-power transmitters and high-gain antennas. Earth stations typically have transmitter powers of the order of a few kilowatts and antennas up to 30 m in diameter with 60 dB gain. The satellite has much more limited power supplies and cannot carry such a large antenna. The satellite transmitter typically has a power output of a few watts and an antenna gain of 20 dB. The ground station must therefore have a very high-gain, low-noise, receiver to enable it to receive signals of the order of a picowatt successfully. Newer satellite systems make use of uplink and downlink channels at 14 and 12 GHz respectively making it possible for higher gain antennas to be used.

Direct broadcasting of television by satellite (DBS) has different requirements depending upon whether the transmissions are received at the central station of a cable TV network or directly by the consumers. The former systems have the same general characteristics as the general-purpose systems described above because it is economically viable for the cable TV operators to make use of an expensive high-gain antenna and low-noise microwave preamplifier. For direct reception in the home the satellite transmitter must have an output power of about 200 W to allow smaller antennas and cheaper preamplifiers to be used. The systems have 12 GHz downlink and 17 GHz uplink frequencies.

12.6 RADAR

Radar (RAdio Detection And Ranging) was the first use to which microwave power sources were put. The first systems operated by transmitting pulses of microwave power from a rotating antenna. Some of the signal was reflected from targets such as aircraft and the time delay of the reflected pulse measured the distance from the transmitter to the target. The direction in which the antenna was pointing supplied information about the direction of the target. From these early beginnings a whole radar family has grown up which includes both pulsed and continuous wave (c.w.) systems. Examples are radars for: threat warning, target location, air traffic control, missile guidance, speed measurement and automatic landing. The information which can be extracted from the returned signal includes: range, velocity, acceleration, angular direction (horizontal and vertical) and target size, shape and identification.

A pulsed radar emits short pulses (a few microseconds) of microwave energy at a fixed pulse repetition frequency (p.r.f.) as shown in Fig. 12.10(a). The returned pulses (of much lower amplitude) resulting from reflection of the first transmitted pulse might be as shown in Fig. 12.10(b). Pulse A has an unambiguous time delay from the transmitted pulse so the range of the target is known. Pulse B has returned during the next transmitted pulse. Unless precautions are taken, the leakage of the transmitted pulse into the receiver would destroy it. For this reason a pulsed radar always incorporates a device to protect the receiver during the transmitter

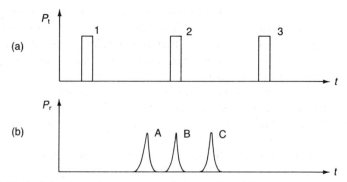

Fig. 12.10 Radar pulses: (a) transmitted pulses and, (b) received pulses. A gives unambigous range information, B is blanked by a transmitter pulse and C gives ambiguous range information.

pulse. This device, which is either a gas discharge 'TR cell' or a semiconductor switch, short circuits the receiver input. Thus pulse B remains undetected. Pulse C has returned after the next transmitted pulse and there is therefore uncertainty about whether it is a reflection of pulse 1 or pulse 2. If unambiguous range information is required it is necessary to use a low p.r.f. so that all possible returns occur before the next pulse.

The ability of a radar to discriminate between two closely spaced targets depends upon the pulse length as illustrated in Fig. 12.11. The long transmitted pulse in Fig. 12.11(a) produces a single long returned pulse. When the pulse length is shortened (Fig. 12.11(b)) it is revealed that there are two targets close together whose returns overlapped when the transmitter pulse was longer. Getting high resolution by shortening the pulse length has the disadvantage that a higher transmitter power may be needed to provide sufficient energy in the returned pulses. An alternative is to use pulse-compression techniques. A pulse-compression radar transmits a long

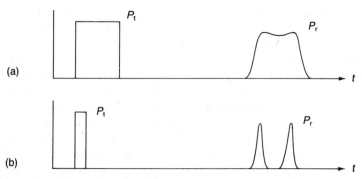

Fig. 12.11 Resolution of a radar: (a) with a long pulse and, (b) with a short pulse.

pulse during which the frequency is swept over a range of a few hundred megahertz as shown in Figs. 12.12(a) and (b). The returned pulse is passed through a dispersive filter which delays the lower frequencies more than the higher ones. The output from this filter is then a much shorter pulse with greater amplitude as shown in Fig. 12.12(c). A pulse-compression radar requires a power amplifier (klystron, TWT or CFA) in place of the magnetron oscillators used in fixed-frequency radar.

The power density at the target is given by (5.46)

$$S_1 = \frac{P_t G}{4\pi R^2},\tag{12.19}$$

where P_t is the power and G the antenna gain of the transmitter, and R is the range. The power reflected by the target depends upon its size and shape and the material from which it is made. These are grouped together as an effective area known as the target cross-section (σ). The power density of the reflected signal at the transmitter is therefore

$$S_2 = \frac{P_t G \sigma}{(4\pi R)^2}.\tag{12.20}$$

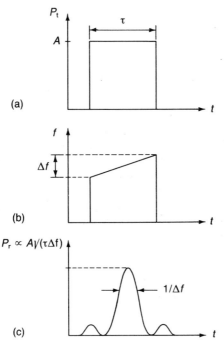

(a)

(b)

(c)

Fig. 12.12 Pulses for pulse-compression radar: (a) transmitted power, (b) transmitted frequency, and (c) compressed received pulse.

If the effective area of the radar antenna is A_e then the power received is

$$P_r = \frac{P_t G \sigma A_e}{(4\pi R)^2}.\tag{12.21}$$

Normally the same antenna is used for transmission and reception so that we can make use of the relationship between the effective area and the gain of an antenna (5.54)

$$A_e = \frac{G\lambda^2}{4\pi}\tag{12.22}$$

to produce an expression for the power received. If the system and propagation losses are represented by L then the result is

$$P_r = \frac{P_t G^2 \lambda^2 \sigma}{(4\pi)^3 R^4 L}.\tag{12.23}$$

If this signal is to be detected reliably then it must exceed the noise power in the receiver. Noise is a manifestation of random thermal events which is spread uniformly over all frequencies. Thus the noise power in the receiver can be written

$$N = FkTB,\tag{12.24}$$

where k is Boltzmann's constant, T the absolute temperature and B the receiver bandwidth. Together these terms represent the background noise radiated from the sky. The constant F known as the noise figure represents the additional noise generated within the receiver. For a fuller discussion of noise in telecommunication systems see Dunlop and Smith (1984). Combining (12.22) and (12.23) yields

$$\frac{P_r}{N} = \frac{P_t G^2 \lambda^2 \sigma}{(4\pi)^3 R^4 L FkTB}.\tag{12.25}$$

This equation is known as the radar equation. The left-hand side is the signal-to-noise ratio and this must exceed a specified level if the system is to function correctly. An important consequence of this equation is that the transmitter power must be multiplied by 16 if the range is to be doubled.

The effective target cross-section decreases rapidly when the target is smaller than one wavelength. For a sphere of radius a where $a \gg \lambda$ the cross-section is πa^2. A flat plate of area A normal to the incident wave has a cross-section

$$\sigma = \frac{4\pi A^2}{\lambda^2}.\tag{12.26}$$

Radars employing short pulses or pulse compression can sometimes discriminate between the cross-sections of different parts of the target. The reflected pulse then has a characteristic shape, known as the *signature* of the target, which can be used to identify it.

One interesting radar cross-section is that of an antenna whose input has been short circuited. The power received is

$$P_r = SA_e \qquad (12.27)$$

so that the power reflected back towards the transmitter is

$$P = SA_e G \qquad (12.28)$$

thus the radar cross-section of the antenna is

$$\sigma = A_e G = \frac{G^2\lambda^2}{4\pi}. \qquad (12.28)$$

This means that the gain and radiation pattern of an antenna can be measured by measuring the signal reflected from it when it is short circuited.

If the target is moving relative to the transmitter then the returned frequency is Doppler shifted. The frequency shift can then be used to measure the relative velocity. This has found an everyday application in the radar speed meters used by the police. Other uses include vehicle detectors for portable traffic lights, aircraft landing systems and docking systems for oil tankers and spacecraft. Doppler systems can use either pulsed or c.w. sources. In pulsed Doppler systems the frequency shift is detected by a set of filters which discriminate between adjacent narrow bands of frequencies. The p.r.f. and its harmonics produce signals which may fall in the range of the filters and there are therefore certain velocities which cannot be measured by a pulsed Doppler system.

Much effort and ingenuity has gone into the processing of radar signals to maximize the ability of systems to detect targets in the presence of reflections from trees, hills, buildings and the like (known collectively as 'clutter') and of signals from other systems. Deliberate attempts to produce misleading responses in a radar system, commonplace in the military environment, are known as electronic countermeasures (ECM) (see Section 12.7). For a fuller discussion of radar systems see the book by Skolnik (1967).

12.7 ELECTRONIC COUNTERMEASURES

The development of ever more advanced military radar systems has been paralleled by the development of electronic countermeasures (ECM) systems. These have the purpose of providing some defence against hostile radars. The threat may be from a surveillance radar or from the radar guidance system of a missile. Simple techniques are the deployment of decoys or of 'chaff' (thin aluminium strips designed to have a large radar cross-section). Efforts are also made to minimize the radar cross-sections of potential targets to make them harder to detect.

More sophisticated ECM systems employ transmitters to produce signals designed to confuse the hostile radar. These signals may be from high-power noise sources or oscillators and designed to swamp the radar reflection with a much bigger jamming signal. If, as often, the frequency of the hostile radar is not known in advance an alternative technique is to use a wide-band receiver and transmitter. This equipment receives the hostile signal and then retransmits it at a higher power level in a way which confuses the hostile radar.

The development of ECM has led to electronic counter countermeasures (ECCM). These techniques are designed to make a radar harder to jam. One common method is to use rapid and random changes in the operating frequency (frequency agility).

12.8 INDUSTRIAL AND DOMESTIC APPLICATIONS

The main industrial application of microwaves is for heating. The applications include cooking, drying, curing and hardening, melting and sterilization. The first of these is familiar in the form of the domestic microwave cooker. The different ways of applying the microwave power fall into two main categories: resonant and non-resonant. In a resonant cavity the field patterns are known and fixed. The material to be processed is usually a lossy dielectric so it must be placed in a region of maximum electric field. Figure 12.13 shows how a cylindrical cavity resonating in the TM_{010} mode can be used for heating a dielectric liquid.

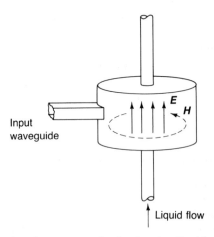

Fig. 12.13 Single-mode microwave cavity for heating liquids.

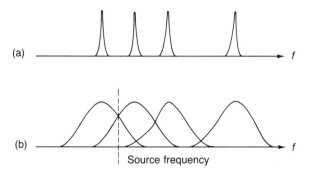

Fig. 12.14 Spectrum of resonances in a multi-mode microwave oven: (a) with the oven empty, and (b) with the widths of the resonances increased by loading.

Single-mode cavities are not suitable for batch heating processes because of the very non-uniform distribution of power within the cavity. The solution is to use a larger cavity so that there are several higher-order modes close to the source frequency. The presence of the dielectric load in the cavity lowers the Q of the resonances causing their resonance curves to overlap as shown in Fig. 12.14. There is then appreciable coupling of the input power into two or more modes each of which has several regions of high electric field within the cavity. To obtain more uniform heating still it is usual either to mount the load on a turntable so that it moves through the field pattern, or else a rotating metal paddle (a 'mode stirrer') is arranged to alter the coupling of the source to the different modes of the cavity. Multi-mode cavities are used in domestic ovens and for a wide range of industrial processes.

Example

A domestic microwave oven which operates at 2.45 GHz has internal dimensions: width 330 mm, depth 338 mm, height 268 mm. Investigate the mode spectrum in the region of the operating frequency.

Solution

The resonant frequencies of a rectangular cavity are given by equation (7.41). Substitution of values of m, n and l gives the range of possible resonances. Finding the complete mode spectrum is time consuming and is best programmed on a computer. Examples of modes close to the source frequency are given in Table 12.1.

The presence of the dielectric load moves the resonances downwards and broadens them. Thus the (3,3,3) resonance could well be excited in a

Table 12.1

m	n	l	f (GHz)
1	4	3	2.483
4	1	3	2.512
3	3	3	2.538

particular case. This mode has 27 regions of high electric field so that the load could be quite uniformly heated by rotating it.

A third way of coupling microwave power into a load is to use some kind of travelling-wave applicator. It may be sufficient to use the fields in a rectangular waveguide as shown in Fig. 12.15. The items to be heated pass through the guide on a conveyor belt or in a pipe like the water calorimeter shown in Fig. 11.3. This arrangement suffers from the disadvantage that the size of the items to be heated is limited by the size of the waveguide. Alternative designs which can heat larger objects employ meander lines and other slow-wave structures.

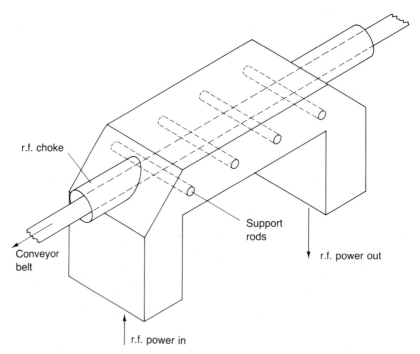

Fig. 12.15 Arrangement of a waveguide applicator for microwave heating.

A full review of microwave heating and drying techniques and their applications is given by Metaxas and Meredith (1983).

12.9 MEDICAL APPLICATIONS

The ability of microwave power to heat biological tissue, exploited in the microwave oven, has a number of consequences in the medical field. It is clearly dangerous for human tissue to be exposed to too much microwave power and anyone working with microwave power systems needs to be aware of the hazards and the precautions which must be taken to ensure safe working. The microwave power heats the body tissue and permanent damage may occur if the heat generated cannot be removed quickly enough by the blood stream. Some parts of the body, especially the eyes are particularly sensitive to microwave radiation. Care must be taken never to look into an open waveguide or antenna even at low power levels. In the USA and Western Europe a figure of $10\,\text{mW}\,\text{cm}^{-2}$ was adopted as the maximum safe exposure on the basis of heating effects. In the Soviet Union and Eastern Europe the level was set at $10\,\mu\text{W}\,\text{cm}^{-2}$ on the basis of non-heating effects which have been observed in laboratory experiments.

In medicine it is common to use heat applied externally to assist healing and to reduce pain. Microwave heating has the advantage that the power can penetrate to the region inside the body where it is needed. Power levels from $100\,\text{mW}\,\text{cm}^{-2}$ to a few watts per square centimetre are used. This technique is known as microwave hyperthemia. The possibility of using it as a way of selectively heating malignant growths is being investigated.

Microwave power is also used in radiotherapy for the treatment of cancer. Electrons accelerated in a linear accelerator (see Section 10.8) are used to produce high-energy X-rays either by collision with a target or by direct bombardment of the human body.

12.10 COMPUTER-AIDED DESIGN OF MICROWAVE SYSTEMS

The complicated interactions between the components of a microwave system make it difficult to design. In particular the possibility of resonances and of effects due to the loading of one component by another need to be carefully investigated. Computer-aided design (CAD) methods are now commonly used in the microwave industry to simulate the performance of systems before they are built. These packages contain routines for carrying out transmission-line manipulations and mathematical models of many of the different kinds of transmission line and component which may be used. The results of calculations can be displayed graphically as frequency response curves or Smith charts. CAD packages of this kind are a powerful way of investigating and optimizing system performance. The effects of possible design changes can be studied and steps taken to get rid of any unwanted characteristics.

12.11 CONCLUSION

This chapter has reviewed briefly some of the systems which make use of electromagnetic radiation, especially those which work at microwave frequencies. The aim has been to emphasize the importance of the material covered earlier in the book by showing how it is applied in major engineering systems. In each case it has only been possible to provide a very brief introduction to give some idea of the scope of the subject and the way in which it is related to electromagnetic theory. The reader is encouraged to consult the books and papers listed in the bibliography to find out more about these subjects.

The author is convinced that familiarity with the material covered in this book is still a vital part of the professional competence of any electronic engineer.

EXERCISES

12.1 A microwave communication link is to be designed to operate over a 20 km range at 14 GHz. The antennas available have 32 dB gain, the anticipated propagation losses are 6 dB and the path gain factor varies between 0.7 and 1.6. If the signal level at the receiver is to be not less than 1 μW what must the transmitter power be?

12.2 Calculate the signal level at the receiver of a direct-broadcasting satellite system if the satellite is in geostationary orbit, the frequency is 12.1 GHz, the transmitter power is 200 W and the antennas have 26 dB and 34 dB gain.

12.3 What is the range of the radio horizon for a radar set in a small boat whose antenna is 5 m above the sea surface?

12.4 Calculate the signal level at the receiver of a radar set operating at 1.12 GHz whose transmitted power is 1 MW if the antenna gain is 38 dB, the target is at a distance of 97 km and the target cross-section is 10 m^2.

12.5 A radar set is operated with 1 μs pulses and a pulse-repetition frequency of 680 s^{-1}. What are the minimum detection range and the range resolution?

References

Akhtarzad, S. and Johns, P.B. (1975) Solution of Maxwell's equations in three space dimensions and time by the TLM method of numerical analysis, *Proc. IEE*, **122**, 1349.

Baden-Fuller, A.J. (1987) *Ferrites at Microwave Frequencies*. Peter Peregrinus, London.

Bailey, M.C. and Crosswell, W.F. (1982) Antennas. In *Electronic Engineers' Handbook* (D.J. Fink and D. Christiansen, eds.). McGraw Hill, New York.

Balston, D.M. (1989) Pan-European Cellular Radio, *Electronics and Communication Engineering Journal*, **1**, 7–13.

Bar-Lev, A. (1979) *Semiconductors and Electronic Devices*. Prentice-Hall International, London.

Barton, D.K. (1982) Radar Principles. In *Electronic Engineers' Handbook* (D.G. Fink and D. Christiansen, eds.). McGraw-Hill, New York.

Bleaney, B.I. and Bleaney, B. (1976) *Electricity and Magnetism*, 3rd edn. Oxford University Press, Oxford.

Bryant, G.H. (1988) *Principles of Microwave Measurement*. Peter Peregrinus, London.

Carter, G.W. (1967) *The Electromagnetic Field in its Engineering Aspects*, 2nd edn. Longmans, London.

Carter, R.G. (1986). *Electromagnetism for Electronic Engineers*. Van Nostrand Reinhold, Wokingham.

Carter, G.W. and Richardson, A. (1972) *Techniques of Circuit Analysis*. Cambridge University Press, Cambridge.

Collin, R.E. (1966) *Foundations for Microwave Engineering*. McGraw-Hill, New York.

Connolly, D.J. (1976) Determination of the interaction impedance of coupled-cavity slow-wave structures, *IEEE Trans. on Electron Devices*, **ED-23**, 491–3.

Dekker, A.J. (1959) *Electrical Engineering Materials*. Prentice Hall, Englewood Cliffs, NJ.

Di Giacomo, J.J. *et al.* (1958) *Design and Fabrication of Nanosecond Digital Equipment*. RCA, Princeton, NJ.

Dunlop, J. and Smith, D.G. (1984) *Telecommunications Engineering*. Van Nostrand Reinhold (UK), Wokingham.

Edwards, T.C. (1981) *Foundations for Microstrip Circuit Design*. Wiley-Interscience, Chichester.

Field, J.C.G. (1983) An introduction to electromagnetic screening theory, *IEE Colloquium Digest No. 1983/88.*

Forrer, M.P. and Jaynes, E.T. (1960) Resonant modes in waveguide windows, *IRE Trans. on Microwave Theory and Techniques*, **MTT-8**, 147–50.

Fox, A.G. (1947) An adjustable waveguide phase changer, *Proc. IRE*, **35**, 1489–98.

Fujisawa, K. (1958) General treatment of klystron resonant cavities, *IRE Trans. on Microwave Theory and Techniques*, **MTT-6**, 344–58.

Gardiol, F.E. (1984) *Introduction to Microwaves*. Artech House, Dedham, Mass.

Getsinger, W.J. (1973) Microstrip dispersion model, *IEEE Trans. on Microwave Theory and Techniques*, **MTT-22**, 34–9.

Gilmour, A.S.Jr (1986) *Microwave Tubes*. Artech House, Dedham, Mass.

Granatstein, V.L. and Alexeff, I. (eds.) (1987) *High Power Microwave Sources*. Artech House, Dedham, Mass.

Green, R.J. (1989) Optical communications; past, present and future, *Electronics and Communication Engineering Journal*, **1**, 105–14.

Halbach, K. and Holsinger, R.F. (1976) SUPERFISH a computer program for the evaluation of RF cavities with cylindrical symmetries, *Particle Accelerators*, **7**, 213–22.

Harvey, A.F. (1963) *Microwave Engineering*. Academic Press, New York.

Jackson, G.A. (1989) Survey of EMC measurement techniques, *Electronics and Communication Engineering Journal*, **1** (2), 61–70.

Jones, E.C.Jr and Hale, H.W. (1982) Filters and attennuators. In *Electronic Engineers' Handbook* (D.G. Fink and D. Christiansen, eds.). McGraw-Hill, New York.

Jordan, E.C. and Balmain, K.G. (1968) *Electromagnetic Waves and Radiating Systems*, 2nd edn. Prentice Hall, Engelwood Cliffs, NJ.

Kajfez, D. and Guillon, P. (eds.). (1986) *Dielectric Resonators*. Artech House, Dedham, Mass.

Keiser, B.E. (1983) *Principles of Electromagnetic Compatibility*. Artech House, Dedham, Mass.

Kirby, R.C. (1982) Radio-wave Propagation. In *Electronic Engineers' Handbook* (D.G. Fink and D. Christiansen, eds.). McGraw-Hill, New York.

Laverghetta, T.S. (1976) *Microwave Measurements and Techniques*. Artech House, Dedham, Mass.

Levy, R. (1959) A guide to the practical application of Chebyshev functions to the design of microwave components, *Proc. IEE*, **106c**, 193–9.

Liao, S.Y. (1980) *Microwave Devices and Circuits*. Prentice-Hall, Engelwood Cliffs, NJ.

Longhurst, R.S. (1973). *Geometrical and Physical Optics*, 3rd edn. Longmans, London.

Louisell, W.H. (1960) *Coupled Mode and Parametric Electronics*. Wiley, New York.

McLachlan, N.W. (1955) *Bessel Functions for Engineers*, 2nd edn. Oxford University Press, Oxford.

Marcuvitz, N. (1986) *Waveguide Handbook*. Peter Peregrinus, London.

Matthaei, G.L., Young, L. and Jones, E.M.T. (1964) *Microwave Filters, Impedance Matching Networks and Coupling Structures*. McGraw-Hill, New York.

Metaxas, A.C. and Meredith, R.J. (1983) *Industrial Microwave Heating*. Peter Peregrinus, London.

O'Reilly, J.J. (1984) *Telecommunication Principles*. Van Nostrand Reinhold, (UK), Wokingham.

Pengelly, R.S. (1986) *Microwave Field Effect Transistors*, 2nd edn. Research Studies Press, Letchworth.

Pierce, J.R. (1954) Coupling of modes of propagation, *Journal of Applied Physics*, **25**, 179.

Ramo, S., Whinnery, J.R. and van Duzer, T. (1965) *Fields and Waves in Communication Electronics*. Wiley, New York.

Rosser, W.G.V. (1964) *An Introduction to the Theory of Relativity*. Butterworths, London.

Rudge, A.W., Milne, K., Olver, A.D. and Knight, P. (eds.) (1982–3). *The Handbook of Antenna Design* (2 vols.). Peter Peregrinus, London.

Saad, T.S. (ed.) (1971) *Microwave Engineers' Handbook*. Artech House, Dedham, Mass.

Schelkunoff, S.A. (1943) *Electromagnetic Waves*, Van Nostrand, New York.

Seely, S. (1972) *An Introduction to Engineering Systems*. Pergamon, New York.

Skolnik, M.J. (1967) *Introduction to Radar Systems*. McGraw-Hill, New York.

Topping, J. (1962). *Errors of Observation and their Treatment*, 3rd edn. Chapman and Hall, London.

Waldron, R.A. (1967) *The Theory of Waveguides and Cavities*. Applied Science Publisher, Baukung.

Watson, J. (1988) *Opto-electronics*. Van Nostrand Reinhold International, London.

Webb, R.T. (1985) Ten years of TWT progress, *Electronics and Power*, **31**, 120.

Weiland, T. (1983) TBCI and URMEL. New computer codes for wake field and cavity mode calculations, *IEEE Trans. on Nuclear Science*, **NS-30**, 2489–91.

Weiland, T. (1985) Computer modelling of two- and three-dimensional cavities, *IEEE Trans. on Nuclear Science*, **NS-32**, 2738.

Transmission lines

This appendix summarizes some of the equations which relate to the propagation of waves on transmission lines.

Any two-wire line which supports TEM waves can be modelled by the equivalent circuit shown in Fig. A1. Analysis of this circuit (Carter, 1986) shows that the voltage and current obey the equations

$$\frac{\partial V}{\partial x} = -L\frac{\partial I}{\partial t} \tag{A.1}$$

and

$$\frac{\partial I}{\partial x} = -C\frac{\partial V}{\partial t}. \tag{A.2}$$

Eliminating the variables produces the wave equations

$$\frac{\partial^2 V}{\partial x^2} = LC\frac{\partial^2 V}{\partial t^2} \tag{A.3}$$

and

$$\frac{\partial^2 I}{\partial x^2} = LC\frac{\partial^2 I}{\partial t^2} \tag{A.4}$$

so that waves propagate as $\exp j(\omega t - kx)$ where

$$k = \pm\omega\sqrt{(LC)}. \tag{A.5}$$

The ratio of the voltage to the current is the characteristic impedance given by

$$Z_0 = \left(\frac{L}{C}\right)^{\frac{1}{2}}. \tag{A.6}$$

If the line is terminated so that an incident wave V_i produces a reflected wave ϱV_i, where ϱ is the reflection coefficient, then

$$V = V_i \exp j(\omega t - kx) + \varrho V_i \exp j(\omega t + kx). \tag{A.7}$$

The amplitude of the voltage at a point on the line x from the termination is given by

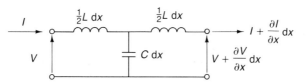

Fig. A1 Equivalent circuit of a two-wire line.

$$|V| = |V_i|\,|1 + \varrho e^{2jkx}| \tag{A.8}$$

which has maximum and minimum values

$$V_{\max} = |V_i|(1 + |\varrho|) \tag{A.9}$$

and
$$V_{\min} = |V_i|(1 - |\varrho|). \tag{A.10}$$

The ratio of these is the voltage standing wave ratio S so that

$$S = \frac{1 + |\varrho|}{1 - |\varrho|} \tag{A.11}$$

and
$$|\varrho| = \frac{S - 1}{S + 1}. \tag{A.12}$$

The apparent impedance at a point on a transmission line is given by the ratio of the voltage to the current at that point. For a line terminated by an impedance Z_L as shown in Fig. A2 the impedance at the plane A–A is given by

$$\frac{Z_L'}{Z_0} = \frac{Z_L + jZ_0\tan kl}{jZ_L\tan kl + Z_0}. \tag{A.13}$$

A quarter-wave section of line transforms impedance so that

$$Z_0^2 = Z_L Z_L'. \tag{A.14}$$

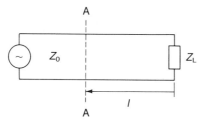

Fig. A2 A line terminated by an impedance.

THE SMITH CHART

A very valuable aid to calculations based on equation (A.14) is the Smith chart. The basis of this chart is a plot of complex reflection coefficient in polar coordinates as shown in Fig. A3. Now as the point of observation is moved along the line away from the termination the only effect is to change the phase so that the tip of the vector in Fig. A3 traces a circle. But the reflection coefficient is related to the impedance by

$$\varrho = \frac{Z - Z_0}{Z + Z_0} \tag{A.15}$$

or
$$\varrho = \frac{z - 1}{z + 1}, \tag{A.16}$$

where z is the impedance normalized to Z_0. When contours of constant impedance are plotted on the polar diagram shown in Fig. A3 the result, the Smith chart, is as shown in Fig. A4. This chart is a graphical representation of equation (A.13). Distance along the transmission line is represented by rotation around the chart with one complete revolution being equivalent to half a wavelength. The standing-wave pattern on a line repeats itself every half wavelength as can be seen from (A.8). Movement towards the generator of the point at which the impedance is measured is represented by clockwise movement around the chart and movement towards the load by anticlockwise rotation.

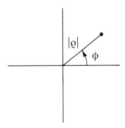

Fig. A3 Plot of complex reflection coefficient in polar coordinates.

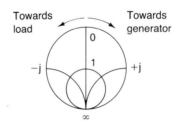

Fig. A4 Graphical representation of equation (A. 13).

Impedances which are wholly real are represented by points on the vertical diameter of the chart. Points on the right-hand half of the chart represent impedances with positive imaginary parts. Impedances with negative imaginary parts are plotted on the left-hand side. Impedances whose value is pure imaginary are plotted around the perimeter of the chart.

There are two points in each revolution of the chart at which the impedance is real. These points correspond to the maxima and minima of the standing wave where

$$z = S \qquad\qquad (A.17)$$

and
$$z = \frac{1}{S}. \qquad\qquad (A.18)$$

This property of the chart enables the impedance of a load at a reference plane to be determined from slotted line measurements. Since the value of S and the position of a standing-wave minimum are known the impedance at that plane can be plotted on the chart using (A.18). The transformation of this impedance to the reference plane as shown in Fig. 11.10 is then performed by moving the appropriate angle around the chart at a constant radius.

The admittance at a point is obtained by reflecting the point representing the impedance in the centre of the chart as shown in Fig. A5. The Smith chart can be used to transform admittances using the procedure outlined above for impedances.

A full discussion of the derivation and use of the Smith chart is given by Dunlop and Smith (1984).

LOSSY LINES

The theory given above is only applicable to lossless lines. A lossy line can be represented by the equivalent circuit shown in Fig. A6. The propagation constant is then

$$k = \pm\omega\sqrt{\left[\left(L + \frac{R}{j\omega}\right)\left(C + \frac{G}{j\omega}\right)\right]} \qquad\qquad (A.19)$$

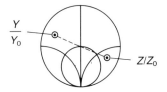

Fig. A5 Obtaining the admittance at a point.

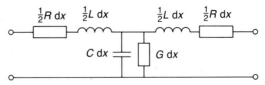

Fig. A6 Equivalent circuit of a lossy line.

by a straightforward extension of (A.5). If the losses are small and dominated by the resistance of the conductors then

$$k = \pm\omega\sqrt{(LC)}\sqrt{\left(1 + \frac{R}{j\omega L}\right)}$$

$$k \simeq \pm\omega\sqrt{(LC)}\left(1 + \frac{1}{2}\frac{R}{j\omega L}\right)$$

$$k \simeq \pm\left[\omega\sqrt{(LC)} - j\frac{R}{2Z_0}\right].$$

A full discussion of propagation on lossy transmission lines will be found in Ramo *et al.* (1965).

Vector formulae

This appendix summarizes the principal vector formulae used in electro-magnetic theory in Cartesian, cylindrical polar and spherical polar coordinates. It also provides a brief review of Bessel functions. The '^' symbol is used to denote unit vectors in the coordinate directions. All the coordinate systems discussed here are orthogonal systems, that is the three coordinate directions at a point are always at right angles to each other.

CARTESIAN COORDINATES

The system of rectangular Cartesian coordinates for describing the position of a point in space is shown in Fig. B1. Note carefully that this is a right-handed set of axes so that rotation from the x direction to the y direction would cause a right-hand thread screw to advance in the z direction. In Cartesian coordinates the vector formulae are:

Gradient
$$\nabla V = \hat{x}\frac{\partial V}{\partial x} + \hat{y}\frac{\partial V}{\partial y} + \hat{z}\frac{\partial V}{\partial z} \qquad (B.1)$$

Divergence
$$\nabla \cdot A = \frac{\partial A_x}{\partial x} + \frac{\partial A_y}{\partial y} + \frac{\partial A_z}{\partial z} \qquad (B.2)$$

Curl
$$\nabla \wedge A = \begin{vmatrix} \hat{x} & \hat{y} & \hat{z} \\ \dfrac{\partial}{\partial x} & \dfrac{\partial}{\partial y} & \dfrac{\partial}{\partial z} \\ A_x & A_y & A_z \end{vmatrix}. \qquad (B.3)$$

The wave equation is

$$\nabla^2 A = \frac{\partial^2 V}{\partial x^2} + \frac{\partial^2 V}{\partial y_2} + \frac{\partial^2 V}{\partial z^2} = \frac{1}{c^2}\frac{\partial^2 V}{\partial t^2} \qquad (B.4)$$

with the general solution

$$V = V_0 \exp j(\omega t - k_x x - k_y y - k_z z) \qquad (B.5)$$

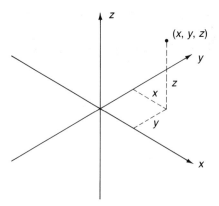

Fig. B1 Rectangular Cartesian coordinate system.

so that

$$k_x^2 + k_y^2 + k_z^2 = \frac{\omega^2}{c^2}.$$ (B.6)

CYLINDRICAL POLAR COORDINATES

The system of cylindrical polar coordinates is shown in Fig. B2. These coordinates are related to rectangular Cartesian coordinates by

$$x = r \cos \theta$$
$$y = r \sin \theta$$
$$z = z.$$ (B.7)

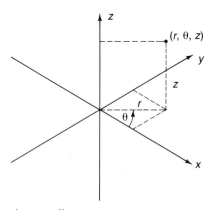

Fig. B2 Cylindrical polar coordinate system.

In cylindrical polar coordinates the vector formulae are:

Gradient
$$\nabla V = \hat{r}\frac{\partial V}{\partial r} + \frac{\hat{\theta}}{r}\frac{\partial V}{\partial \theta} + \hat{z}\frac{\partial V}{\partial z} \tag{B.8}$$

Divergence
$$\nabla \cdot A = \frac{1}{r}\frac{\partial r}{\partial}(rA_r) + \frac{1}{r}\frac{\partial A_\theta}{\partial \theta} + \frac{\partial A_z}{\partial z} \tag{B.9}$$

Curl
$$\nabla \wedge A = \frac{1}{r}\begin{vmatrix} \hat{r} & r\hat{\theta} & \hat{z} \\ \dfrac{\partial}{\partial r} & \dfrac{\partial}{\partial \theta} & \dfrac{\partial}{\partial z} \\ A_r & rA_\theta & A_z \end{vmatrix}. \tag{B.10}$$

The wave equation is

$$\frac{1}{r}\frac{\partial}{\partial r}\left(r\frac{\partial V}{\partial r}\right) + \frac{1}{r^2}\frac{\partial^2 V}{\partial \theta^2} + \frac{\partial^2 V}{\partial z^2} = \frac{1}{c^2}\frac{\partial^2 V}{\partial t^2} \tag{B.11}$$

with the general solution

$$V = V_0\begin{bmatrix} J_n(k_r r) \\ Y_n(k_r r) \end{bmatrix} \exp j(\omega t - n\theta - k_z z), \tag{B.12}$$

where $J_n(kr)$ and $Y_n(kr)$ are known as Bessel functions of order n of the first and second kind respectively. They are analogous to the sines and cosines which appear in solutions to the wave equation in rectangular coordinates and their values can be looked up in tables. Figure B3 shows

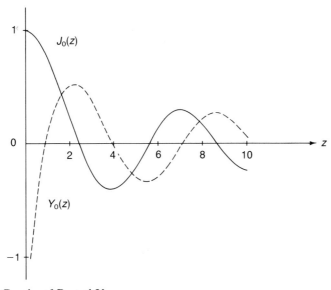

Fig. B3 Graphs of B_0 and Y_0.

graphs of the functions B_0 and Y_0 which occur in problems with cylindrical symmetry. A useful source of information on this subject is McLachlan (1955).

SPHERICAL POLAR COORDINATES

The system of spherical polar coordinates is shown in Fig. B4. They are related to Cartesian coordinates by the equations

$$x = r \sin \theta \cos \phi$$
$$y = r \sin \theta \sin \phi$$
$$z = r \cos \theta. \tag{B.13}$$

In this system of coordinates the vector formulae are:

Gradient
$$\nabla V = \hat{r}\frac{\partial V}{\partial r} + \frac{\hat{\theta}}{r}\frac{\partial V}{\partial \theta} + \frac{\hat{\phi}}{r \sin \theta}\frac{\partial V}{\partial \phi} \tag{B.14}$$

Divergence
$$\nabla \cdot A = \frac{1}{r^2}\frac{\partial}{\partial r}(r^2 A_r) + \frac{1}{r \sin \theta}\frac{\partial}{\partial \theta}(\sin \theta \, A_\theta)$$
$$+ \frac{1}{r \sin \theta}\frac{\partial A_\phi}{\partial \phi} \tag{B.15}$$

Curl
$$\nabla \wedge A = \frac{1}{r^2 \sin \theta}\begin{vmatrix} \hat{r} & r\hat{\theta} & r \sin \theta \, \hat{\phi} \\ \frac{\partial}{\partial r} & \frac{\partial}{\partial \theta} & \frac{\partial}{\partial \phi} \\ A_r & rA_\theta & r \sin \theta \, A_\phi \end{vmatrix}. \tag{B.16}$$

Solutions of the wave equation in spherical polar coordinates are not used in this book so they are not included here.

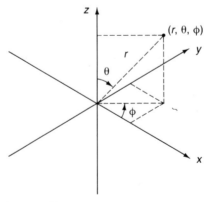

Fig. B4 Spherical polar coordinate system.

Constants and properties of materials

Table C1 *Physical constants*

Primary electric constant (ε_0)	$8.854 \times 10^{-12}\,\mathrm{F\,m^{-1}}$
Primary magnetic constant (μ_0)	$4\pi \times 10^{-7}\,\mathrm{H\,m^{-1}}$
Velocity of light in vacuum (c)	$0.2998 \times 10^9\,\mathrm{m\,s^{-1}}$
Wave impedance of free space (Z_0)	$376.7\,\Omega$
Charge on the electron (e)	$-1.602 \times 10^{-19}\,\mathrm{C}$
Rest mass of the electron (m_0)	$9.108 \times 10^{-11}\,\mathrm{kg}$
Charge/mass ratio of the electron (η)	$1.759 \times 10^{11}\,\mathrm{C\,kg^{-1}}$

Table C2 Properties of dielectric materials

	ε_r	tan δ ($\times 10^4$)
Alumina 99.5%	10	1
Alumina 96%	9	6
Barium titanate	1200	
Beryllia	6.6	1
Epoxy resin	3.5	200
Ferrites	13–16	2
Fused quartz	3.8	1
GaAs (high resistivity)	13	6
Nylon	3.1	200
Paraffin wax	2.25	2
Perspex	2.6	70
Polystyrene	2.54	1.6–2.5
Polystyrene foam	1.05	0.3
Polythene	2.25	3
PTFE (Teflon)	2.08	3.7

Note: the values in this table are typical of those at microwave frequencies. Actual samples of material may have properties which differ from those given here. In some cases they vary appreciably with frequency.

Table C3 Properties of conductors

	Conductivity $(S\,m^{-1})$
Aluminium	3.5×10^7
Animal body tissue (average)	0.2
Brass	1.1×10^7
Copper	5.7×10^7
Distilled water	2×10^{-4}
Ferrite (typical)	10^{-2}
Fresh water	10^{-3}
Gold	4.1×10^7
Iron	0.97×10^7
Nickel	1.28×10^7
Sea water	4
Silver	6.1×10^7
Steel	0.57×10^7

Table C4 Properties of ferromagnetic materials

	μ_r	B_{sat} (T)
Feroxcube 3	1500	0.2
Mild steel	2000	1.4
Mumetal	80 000	0.8
Nickel	600	
Silicon iron	7000	1.3

Answers to selected problems

1 ELECTROMAGNETIC WAVES

1.1 Polystyrene, $229\,\Omega$; alumina, $126\,\Omega$; barium strontium titanate, $3.8\,\Omega$

1.2 $21.8\,\mathrm{W\,m^{-2}}$; $39.7\,\mathrm{W\,m^{-2}}$; $1.32\,\mathrm{kW\,m^{-2}}$

1.3 Silver: $9.1\,\mathrm{mm}$, $28.8\,\mu\mathrm{m}$, $0.91\,\mu\mathrm{m}$
   Graphite: $0.225\,\mathrm{m}$, $0.71\,\mathrm{mm}$, $22.5\,\mu\mathrm{m}$
   Sea water: $35.6\,\mathrm{m}$, $0.11\,\mathrm{m}$, $3.6\,\mathrm{mm}$

1.4 Glass, $3.82\,\mathrm{dB\,m^{-1}}$; fused quartz, $0.177\,\mathrm{dB\,m^{-1}}$

1.5 $8.98\,\mathrm{MHz}$, $898\,\mathrm{MHz}$, $0.148\,\mathrm{MHz}$, $14.8\,\mathrm{MHz}$

1.6 $377\,\Omega$, $166\,\Omega$

1.7 $1.4\,\mathrm{GHz}$, $2.8\,\mathrm{GHz}$, $5.6\,\mathrm{GHz}$

1.8 $k_+ = 26.4\,\mathrm{j}$, $k_- = 10.8\,\mathrm{j}$

2 WAVEGUIDES GUIDED BY PERFECTLY CONDUCTING BOUNDARIES

2.1 (1) $5.24\,\mathrm{mm}$; (2) $5.44\,\mathrm{mm}$; (3) $0.66\,\mathrm{mm}$

2.2 Cut-off wavelength, $144.3\,\mathrm{mm}$;
   guide wavelengths, $215.7\,\mathrm{mm}$, $138.5\,\mathrm{mm}$, $106.4\,\mathrm{mm}$, $87.7\,\mathrm{mm}$;
   phase velocity, $0.414 \times 10^9\,\mathrm{m\,s^{-1}}$
   group velocity, $0.219 \times 10^9\,\mathrm{m\,s^{-1}}$

2.4 $72.1\,\mathrm{mm}$, $30.8\,\mathrm{mm}$, $30.8\,\mathrm{mm}$

2.5 At $13\,\mathrm{GHz}$: characteristic impedance $140\,\Omega$, $279\,\Omega$, $419\,\Omega$, $551\,\Omega$

2.6 Cut-off wavelength $95.1\,\mathrm{mm}$ (air-filled), $142.7\,\mathrm{mm}$ (wax-filled);
   characteristic impedance (at $5\,\mathrm{GHz}$) $453\,\Omega$ (air-filled), $302\,\Omega$ (wax-filled)

3 WAVES WITH DIELECTRIC BOUNDARIES

3.1 $32.3°$

3.2 1.76

3.3 68°, 62°, 58°
3.4 5.6 mm, 4.6 mm
3.5 14.1 mm, 0.246 (12.2 dB return loss)
3.6 9.3 mm, 0.198 (14.1 dB return loss)
3.7 widths: 7.7 mm, 3.4 mm, 0.45 mm

4 WAVES WITH IMPERFECTLY CONDUCTING BOUNDARIES

4.1 Skin depth
    brass 10.7, 3.4, 1.1 μm
    gold 5.6, 1.8, 0.56 μm
    Surface resistance:
    brass 8.5, 26.7, 82.6 mΩ m^{-1}
    gold 0.4, 13.6, 43.6 mΩ m^{-1}
4.2 Transmission loss without film 0.0014 dB; with film 99 dB
4.3 Typical values:
    at 10 Hz, $S_M = 0$ dB, $S_E = 285$ dB
    at 100 MHz, $S_M = 343$ dB, $S_E = 393$ dB
4.4 0.11 dB m^{-1}

5 ANTENNAS

5.1 2.86 m^2, 0.045 m^2, 584 mm^2
5.2 0.24×10^{-9} Ω
5.3 6.56, 3.28, 6.56
5.4 6.40, 25.6, 102.4
5.5 Nulls at:
    5 GHz; 36.9°
    10 GHz; 17.5°, 36.9°, 64.2°
    60 GHz, 2.87°, 5.73°, 8.60° etc.
5.6 0.218 of a wavelength from the centre
    0.038 of a wavelength from the end
    (Note the slot needs to be slightly lengthened)

6 COUPLING BETWEEN WAVE-GUIDING SYSTEMS

6.1 1.68, 1.06
6.2 S reduced by a further 6 dB
6.3 Waveguide height 1.61 mm;
    step height 4.41 mm, step length 17.95 mm;
    short circuit 17.95 mm behind the junction
6.4 $C = 177$ pF/m^{-1}, $L = 0.63$ mH m^{-1};
    $C' = 9$ pF/m^{-1}, $M = 0.03$ mH m^{-1}
    Impedances: 61 Ω, 55 Ω
    $k = 61.1$ m^{-1}

7 ELECTROMAGNETIC RESONATORS AND FILTERS

7.1 $L = 21.1\,\text{nH}$; $C = 0.104\,\text{pF}$, $R = 270\,\text{k}\Omega$;
   $37.7\,\text{k}\Omega$, $82.0°$;
   $25.5\,\text{k}\Omega$, $-84.6°$
7.2 4.85, 9.71, 14.56 MHz
7.3 21.4, 159 MHz
7.4 11.34, 14.39, 15.29 GHz;
   6.06, 8.18, 7.70 GHz
7.5 101 dB
7.6 $r = 0.56\,\text{mm}$; $h = 0.28\,\text{mm}$

8 FERRITE DEVICES

8.1 Typical values:
   $2.36 - 0.009\text{j}$, $2.07 - 0.0057\text{j}$ at 1 GHz
   $13.7 - 13.2\text{j}$, $1.61 - 0.015\text{j}$ at 8 GHz
   $-4.76 - 1.82\text{j}$, $1.55 - 0.015\text{j}$ at 10 GHz
8.2 Typical values:
   $124.7 - 0.24\text{j}$, $116.8 - 0.16\text{j}$ at 1 GHz
   $2626 - 1059\text{j}$, $825 - 3.8\text{j}$ at 8 GHz
   $333 - 1802\text{j}$, $1011 - 4.9\text{j}$ at 10 GHz
8.3 Answers at 1 GHz intervals:
   $y = 6.99, 5.94, 5.20, 4.65, 4.21\,\text{mm}$

10 VACUUM DEVICES

10.1 (1) $41.6 \times 10^6\,\text{m s}^{-1}$; $25.5\,\text{kA m}^{-2}$, 555 MHz;
    (2) $58.5 \times 10^6\,\text{m s}^{-1}$, $25.5\,\text{kA m}^{-2}$, 468 MHz;
    (3) $133.8 \times 10^6\,\text{m s}^{-1}$, $239\,\text{kA m}^{-2}$, 948 MHz
10.2 (1) 75 mm
       13.9, 4.16, 0.69 mm
       92.5, 27.8, 46.3 kΩ
    (2) 125 mm
       19.5, 5.85, 0.98 mm
       6.24, 1.87, 0.31 kΩ
    (3) 141 mm
       44.6, 13.4, 2.23 mm
       3.16, 0.95, 0.16 kΩ
10.3 (1) 100 W, 4.5 W
    (2) 5 kW, 267 W
    (3) 720 kW, 19 kW
10.4 13.6 dB, 114 MHz
10.5 7.6°, 27.9 dB

11 MICROWAVE MEASUREMENTS

11.1 8.059, 9.196, 11.80, 15.61 GHz
11.2 10.39, 11.29, 13.50, 16.94 GHz
11.3 ±0.25 dB
11.4 $0.935 + 0.002j$
     $0.89 - 0.13j$
     $1.23 - 0.32j$
11.5 64.3
11.6 10.0

12 SYSTEMS USING ELECTROMAGNETIC WAVES

12.1 400 W
12.2 0.6 nW
12.3 8 km (9.2 km if atmospheric refraction is included)
12.4 0.16 nW
12.5 150 m

Index